Revitalizing Inclusive Socioeconomics

The Role of SocioEcTech Justice on Human Growth

Aaron W. Wemple

Life Winks LLC

Contents

Blurbs

"Revitalizing Inclusive Socioeconomics" is a thought-provoking and inspiring guide to creating a more equitable and inclusive society. Through a series of engaging chapters, readers will delve into the concepts of socioectech justice, accessibility, and adaptability, exploring how these critical elements impact human growth and development. With real-life case studies of individuals who have overcome socio-economic barriers, this book highlights the importance of access and adaptability in creating a fairer and more just world. Whether you're an educator, a policy maker, or simply someone who wants to make a positive impact on the world, "Revitalizing Inclusive Socioeconomics" is a must-read. With its inspiring stories and insightful analysis, this book is the ultimate resource for anyone looking to revitalize inclusive socioeconomics and promote human growth and development. " - CEO Olivia Sanders

This book is dedicated to all those who strive to make the world a better place, to those who believe in the power of technology to change lives, and to those who never give up in the face of adversity. Your courage, passion, and unwavering commitment to justice inspire us all to work towards a brighter future. May this book serve as a testament to your unwavering spirit, and may it bring hope and encouragement to those who seek a world where technology is used for the betterment of all humanity.

INTRODUCTION

In this book, "Revitalizing Inclusive Socioeconomics: The Role of SocioEcTech Justice on Human Growth," we delve into the critical and urgent need for a more equitable and sustainable world. The rapid pace of technological advancement and its impact on society demands a rethinking of traditional socio-economic models and the role of technology in promoting justice and growth. In today's world, the challenge is to create inclusive socio-economic systems that ensure that all individuals, regardless of their background, have equal access to the benefits and opportunities of technology. To achieve this, we need to look beyond traditional models and embrace a new way of thinking about the interconnections between technology, society, and the environment. This book explores the concept of socioectech justice and its role in promoting inclusivity, sustainability, and resilience in our communities. Through a combination of case studies, expert insights, and a call to action, this book provides a roadmap for building a more just and inclusive tech-enabled society. The future is in our hands, and it is up to us to make a difference.

"Where technology thrives, so too shall we, But let's not forget, fairness must be, For growth and prosperity to be shared, Socioectech justice must be cared."

1

Sarah and Bob's Tale of Two Bakeries

Once, there were two bakeries in two different parts of the world. One was a traditional brick-and-mortar bakery owned by a third-generation baker named Bob, and the other was a small suburban bakery owned by a woman named Sarah.

Bob's bakery had been a staple in the community for over 50 years. Bob learned the art of baking from his grandfather, who opened the bakery and took over the business after his retirement. He was proud of his family's legacy and the reputation the bakery had built over the years for its delicious baked goods and friendly service. However, Bob was facing new challenges in today's fast-paced world. With the rise of big chain bakeries and the increasing popularity of online shopping, his business was losing customers every day. Bob was worried about the future of his bakery and the family legacy that he had been tasked to preserve. He was determined to find a way to keep his business relevant and competitive, but he didn't know where to start.

On the other hand, Sarah was the owner of a small suburban bakery. She was struggling to keep up with the competition and decided to embrace technology to reach a wider audience. With the help of a local tech enthusiast, Sarah transformed her bakery into an online hub for baked goods, allowing customers to place orders and have their treats delivered right to their door. Sarah's business took off, and she quickly became known as the go-to bakery for delicious and convenient treats. Sarah's success showed the importance of embracing technology in today's fast-paced world and how it can provide opportunities for growth and prosperity.

Sarah's success also highlighted the importance of inclusive socioeconomics. By making her business accessible to everyone, regardless of their location or financial status, Sarah ensured that her bakery was a place for everyone to enjoy. This is the essence of inclusive socioeconomics, a principle that will be explored in depth in the following chapters.

The tale of these two bakeries shows the importance of both technology and inclusive socioeconomics in promoting human growth and prosperity. Sarah's story is just one example of how these principles can be applied in practice, and we will continue to examine their impact on human growth and development throughout the book. So, grab a cup of coffee and join us on this exciting journey as we delve into the world of revitalizing inclusive socioeconomics!

2

Tale of Two Cities: The Impact of SocioEcTech Justice

In the heart of America, there lie two cities that stand in stark contrast to one another. The first city is a hub of technology and innovation, where businesses thrive and people live lives of comfort and convenience. The second city, on the other hand, is a place where technology and innovation are largely inaccessible. This city is plagued by poverty, unemployment, and a lack of opportunities. In this chapter, we will explore the impact of socio-economic technology justice on these two cities and the lives of their residents.

The tech-savvy city

The first city is a place where technology is integrated into every aspect of life. Here, people have access to the latest technology and enjoy the convenience of a modern lifestyle. The businesses in this city have embraced technology and have reaped the benefits. From online shopping to virtual classrooms, technology has transformed the way they operate.

The thriving bakery in a different city

Meet Sarah, the owner of a small bakery in this city. Sarah was struggling to keep up with the competition and was looking for a way to reach a wider audience. With the help of a local tech enthusiast, Sarah transformed her bakery into an online hub for baked goods, allowing customers to place orders and have their treats delivered right to their door. Sarah's business took off, and she quickly became known as the go-to bakery for delicious and convenient treats.

Sarah's success highlights the importance of embracing technology in today's fast-paced world. By using technology to reach a wider audience and make her business more accessible, Sarah was able to overcome the challenges facing traditional bakeries and find success. This is a prime example of how technology can be used to create opportunities for growth and prosperity.

The tech-deprived City

The second city is a place where technology is largely inaccessible. The residents here face a stark reality, one where they are unable to compete in today's job market or take advantage of the convenience and comfort that technology provides. Unemployment and poverty are rampant, and the lack of opportunities for growth and prosperity is a constant burden.

The struggling bakery in a tech-starved city

Meet Bob, owner of a traditional brick-and-mortar bakery in this city. Despite being in business for many years, Bob's bakery is struggling to keep up with the competition. With the rise of big chain bakeries and the increasing popularity of online shopping, Bob's business is losing customers every day. Bob is worried about the future of his bakery and the family legacy that he has been tasked to preserve.

Bob's story illustrates the challenges faced by many traditional businesses in today's economy, particularly in tech-deprived areas. The lack of access to technology has left Bob and his business at a disadvantage, unable to compete with businesses in more technologically advanced cities.

The importance of revitalizing inclusive socioectech justice

The difference between these two bakeries and the experiences of their residents highlights the importance of socio-economic justice. In the tech-savvy city, businesses like Sarah's bakery have been able to thrive and provide opportunities for growth and prosperity. Meanwhile, in the tech-deprived city, businesses like Bob's bakery struggle to survive and provide opportunities for their residents.

This disparity is unacceptable, and we must work towards creating a world where everyone has access to the tools and opportunities necessary for success. This is the core principle of socio-economic technology justice, and it is the foundation of the discussion that follows in this book. Through exploring these two cities and the experiences of their residents, we can see the impact that socio-economics have.

The difference between the two cities and the experiences of Sarah and Bob highlights the importance of revitalizing inclusive socioectech justice. In one city, businesses are thriving and providing opportunities for growth and prosperity, while in the other, businesses are struggling to survive. This disparity is unacceptable, and we must work towards creating a world where everyone has access to the tools and opportunities necessary for success.

By embracing the principles of socio-economic technology justice, we can create a world where everyone has equal opportunities for growth and prosperity. Read on as we take on this important challenge and make a difference for future generations.

3

A Tale of Two Cities: The Impact of SocioEcTech Justice

In today's fast-paced world, technology has become an integral part of our lives, from the way we communicate to the way we do business. While technology has brought many benefits to our society, it has also brought new challenges and inequalities. This is where the concept of socioectech justice comes into play.

Socioectech justice refers to the fair and equitable distribution of technology and its benefits to all members of society, regardless of their background or financial status. In other words, it is about ensuring that everyone has access to the same opportunities and benefits that technology provides.

To understand the impact of socioectech justice, we will look at two cities, each with its own unique approach to technology and its integration into society.

The tech hub

City A is a bustling metropolis known for its cutting-edge technology and innovative start-ups. The city government has invested heavily in technology infrastructure, attracting tech companies and entrepreneurs from around the world. The city is home to the latest gadgets, high-speed internet, and state-of-the-art technology.

But not everyone in City A is benefiting from this tech boom. Many low-income residents and marginalized communities are left behind, unable to afford the latest technology or access the high-speed internet necessary to participate in the digital economy. This has created a widening gap between the haves and the have-nots, with those who are already disadvantaged falling further behind.

The inclusive city

City B is a smaller, more traditional city that has embraced a different approach to technology and its integration into society. The city government has prioritized socioectech justice, working to ensure that everyone has access to the benefits of technology, regardless of their background or financial status.

City B has implemented a number of initiatives to make technology more accessible, such as offering free Wi-Fi in public spaces and providing technology training for residents. These efforts have paid off, with the city's residents enjoying a higher standard of living, better job opportunities, and a more inclusive and prosperous community.

The impact of socioectech justice is clear in these two cities. In City A, where technology is seen as a commodity and is not accessible to all, it has created division and unequal opportunities. In contrast, in City B, where technology is viewed as a tool for human development and is accessible to all, it has brought about growth, prosperity, and a more inclusive community.

The stark contrast between these two cities in the previous chapter, or two counties, states or businesses serves as a poignant reminder of the vital role that socioecotech justice plays in fostering human growth and advancement. The advancements in technology hold immense promise for our society, but it is imperative that we ensure these advancements are not just restricted to a privileged few. Instead, we must strive to create an inclusive and equitable world where everyone, regardless of their socio-economic status, has access to the same technological opportunities and benefits. By embracing the principles of socioecotech justice, we can build a brighter future for all and unleash the full potential of technology to uplift and empower our communities.

As we have seen, the impacts of socioectech justice on two cities can vary greatly. In one city, it can bring about a sense of freedom and empowerment, enabling individuals to access technology and opportunities for growth and prosperity. In another city, it can perpetuate a digital divide, perpetuating inequality and hindering human potential. However, there is hope for a more inclusive and equitable future. By understanding the fundamentals of inclusive socioeconomics, we can work towards a world where everyone has the opportunity to thrive, regardless of their background or circumstances. Let us now delve into the background and principles of inclusive socioeconomics, and see how it lays the foundation for socioectech justice.

4

Understanding the Fundamentals of Inclusive Socioeconomics

Socioecomics, a term coined from combining the words "society" and "economics," refers to the study of how the economic and social aspects of a society interact and affect each other. Inclusive socioeconomics, on the other hand, focuses on ensuring that all members of society have equal access to opportunities and resources that contribute to their overall well-being and growth. This concept is particularly relevant in today's world, where technology and digital advancements have created an ever-growing digital divide between those who have access to technology and those who do not.

The digital divide refers to the unequal distribution of technology, including access to computers, the internet, and digital devices, among different populations. This divide perpetuates social and economic inequalities, hindering the growth and development of those who do not have access to technology. As technology continues to play a crucial role in shaping our world, it is vital to address the digital divide and ensure that all members of society have equal access to technology and the opportunities it provides.

Inclusive socioeconomics, in addition to addressing the digital divide, also involves ensuring that all members of society have equal access to economic opportunities, education, and social services. This includes promoting diversity and inclusivity in the workforce, providing access to affordable housing and healthcare, and advocating for policies that support sustainable and equitable economic growth.

In order for inclusive socioeconomics to be successful, it must be implemented through a multi-faceted approach that takes into consideration the social, economic, and technological aspects of society. This includes, but is not limited to, providing access to technology, fostering diversity and inclusivity in the workforce, advocating for ethical technology regulations, and promoting sustainable economic growth.

The Digital Empowerment Movement in India

The Digital Empowerment Movement in India is a prime example of the importance of inclusive socioeconomics. This movement aims to bridge the digital divide and provide access to technology for those who have been left behind. By setting up community technology hubs in rural areas, the movement provides access to computers, the internet, and digital devices to those who would otherwise not have access.

In addition to providing access to technology, the movement also focuses on digital literacy and skills training. This includes teaching individuals how to use technology to improve their lives, such as by using the internet to find job opportunities or access healthcare information.

The impact of the Digital Empowerment Movement has been significant, with individuals reporting increased economic opportunities and improved quality of life. For example, farmers in rural areas have been able to access information about market prices, leading to increased profits. Additionally, individuals have been able to access healthcare information and connect with healthcare providers, leading to improved health outcomes.

The rise of social entrepreneurship

Social entrepreneurship is another important aspect of inclusive socioeconomics. Social entrepreneurs are individuals who start businesses with the goal of creating positive social and environmental change, in addition to generating profits.

One example of a successful social enterprise is Grameen Bank, founded by Muhammad Yunus in Bangladesh. Grameen Bank provides microfinance loans to individuals who would otherwise not have access to financial services. This has led to improved economic opportunities and reduced poverty in Bangladesh.

Another example is the mobile phone provider, Grameenphone, which has provided access to mobile phones and internet services to individuals in rural areas of Bangladesh. This has led to improved access to information and communication, leading to increased economic opportunities and improved quality of life.

Mrs. Smith: An elderly victim of cybercrime

Mrs. Smith was a 72-year-old retiree living in a small town in the United States. She lived a quiet life, spending her days gardening and reading books. She was well-known in her community as a kind and friendly person who was always willing to help others.

However, Mrs. Smith's peaceful existence was shattered when she became the victim of cybercrime. A neighbor, who had a grudge against her, accused her of stealing money from a local charity event. The case was brought to court, and the hearing was conducted virtually due to the ongoing pandemic.

Unfortunately, Mrs. Smith did not have a computer and was unable to attend the hearing. She was found guilty in her absence and was ordered to pay a large fine. The ruling was devastating for Mrs. Smith, who was living on a fixed income and could not afford to pay the fine.

The case of Mrs. Smith highlights the importance of ensuring that all citizens have equal access to technology, regardless of their age or income. This is a prime example of how the digital divide can have serious consequences for vulnerable populations, especially during a pandemic when many court proceedings are being conducted online.

Furthermore, the case also highlights the need for better regulation and oversight of virtual court proceedings. It is imperative that all parties involved in these proceedings have access to the necessary technology and support to participate effectively. This includes ensuring that the elderly and low-income individuals are not left behind in the digital world and are not unfairly punished because of their lack of technology access.

In conclusion, the case of Mrs. Smith highlights the importance of ensuring that all citizens have equal access to technology, and that virtual court proceedings are conducted in a fair and equitable manner. This is an important aspect of inclusive socioeconomics, which aims to ensure that everyone has the opportunity to participate in the digital economy and have a fair shot at prosperity. By working together to promote access to technology and equitable virtual court proceedings, we can help ensure that no one is left behind in our increasingly digital world.

In conclusion, the impact of socioectech justice cannot be overstated. It is the foundation for creating a more inclusive and equitable world for all, regardless of background or circumstances. The principles of accessibility, diversity and inclusion, adaptability, and transparency and accountability are the cornerstones of a just and thriving society. The rise of social entrepreneurship, community technology hubs, and the digital empowerment movement are just some of the examples of how socioectech justice is being put into practice and creating positive change.

However, there are still many challenges and barriers to overcome in order to ensure that everyone has access to the benefits of technology and the opportunities it provides. The future of inclusive socioeconomics is contingent on the ability of society to address these issues and promote sustainable growth. We must work together to build a movement for change, so that the next generation can inherit a world that is not only technologically advanced, but also just and equitable.

In this book, we have explored the story of two cities and how socioectech justice has impacted their growth. We have also delved into the fundamentals of inclusive socioeconomics and why it is so crucial for our future. We call on all individuals, organizations, and governments to join us in this movement for change. By working together, we can create a world where freedom, unity, and prosperity are available to everyone. Let us all do our part to make this a reality.

5

Why SocioEcTech Justice Matters for Human Growth and Development

In a world that is rapidly advancing technologically, it is important to consider the impact that these advancements will have on society as a whole. With the rise of the digital age, the way that we interact, communicate and live our lives is changing at a breakneck pace. However, not everyone is able to access these technologies, or use them to their full potential, and this is where the concept of socioectech justice comes into play. This chapter will explore why socioectech justice is so important for human growth and development, and how it is essential for building a society that is inclusive, equitable and just.

Socioectech justice is about ensuring that everyone has access to the technologies and resources they need to live their lives to the fullest. For example, imagine an elderly woman living in a remote rural area who is the victim of a crime. She is unable to attend the hearing because she does not have access to a computer, and the hearing is being held virtually due to the ongoing pandemic. She is found guilty of the crime in her absence, even though she is actually the victim. This scenario highlights the importance of socioectech justice, and how it is essential for ensuring that everyone has access to the technologies they need to participate fully in society.

Mrs. Smith's story is a powerful example of the impact of the socioecotech divide on human growth and development. This divide refers to the unequal access to technology and the resulting disparities in opportunity, empowerment, and well-being. In Mrs. Smith's case, the divide was a barrier to her access to justice and the ability to defend herself against false accusations.

One day, Mrs. Smith was approached by a peer group at her local library. They were a group of tech-savvy individuals who were passionate about bridging the socioecotech divide and empowering others to participate in the digital world. They saw that Mrs. Smith was struggling to navigate the virtual hearing process and offered to help.

Together, they walked through the software requirements and showed Mrs. Smith how to access the necessary technology. They also taught her how to communicate effectively with the circuit clerk's office and ensured that her voice was heard in the virtual hearing. Through their guidance and support, Mrs. Smith was able to present her case and ultimately clear her name.

This case highlights the importance of socioecotech justice in promoting human growth and development. By providing access to technology and the skills needed to use it effectively, Mrs. Smith was able to overcome the barriers to justice and have her voice heard. This empowerment allowed her to stand up for herself and reclaim her dignity, something that would have been impossible without the support of her peers.

It also highlights the need for transparency and accountability in the technology sector. By ensuring that everyone has access to the same tools and opportunities, we can create a level playing field where everyone has a chance to succeed. By promoting diversity and inclusion in the technology sector, we can create a more equitable world where everyone has the chance to grow and reach their full potential.

Another example is the case of a young student named Olivia who is passionate about computer programming, but who lives in an inner-city area where technology is scarce. The student is unable to attend a coding boot camp because they do not have access to a computer, and they are unable to find one in their community. Without access to the technology and resources they need, this student will never have the opportunity to pursue their passion and reach their full potential.

In these cases, and many others like them, the absence of socioectech justice has a real and tangible impact on human growth and development. It stifles opportunities, limits potential and perpetuates inequality. This is why socioectech justice matters so much. By ensuring that everyone has access to the technologies and resources they need, we can build a society that is more inclusive, equitable and just.

Meet John Doe, a 68-year-old retiree who lived a simple life in a small town. He lived in a modest house that he had inherited from his parents and spent most of his days gardening and taking care of his pet cat. John was a friendly neighbor, always willing to lend a helping hand.

However, one day, everything changed for John. A neighbor who had a grudge against him, accused John of theft. The case was heard online due to the pandemic and John was unable to attend the hearing in person. He didn't have access to a computer or the internet and was unaware of the virtual hearing.

As a result, John was found guilty in absentia and was sentenced to six months in prison. He was devastated when he found out and couldn't believe that he was being punished for a crime he didn't commit.

John's case was a perfect example of the digital divide and the impact it has on the most vulnerable members of society. He was a victim of circumstance, unable to defend himself because of his lack of access to technology. This case highlights the importance of socioectech justice and the need for accessibility and inclusion in our technological advancements.

Socioectech justice is not just a concept but a necessity for everyone to have equal opportunities and access to technology. Without it, people like John will continue to be left behind and suffer the consequences of a world that is becoming increasingly reliant on technology. It is imperative that we take action now to ensure that no one is left behind and that everyone has equal access to technology and its benefits.

However, socioectech justice is not just about access to technology. It also encompasses other key principles, such as diversity and inclusion, adaptability, and transparency and accountability. These principles help to create a society where everyone is able to participate and thrive, regardless of their background or circumstances.

One of the key ways that socioectech justice is promoted is through the creation of community technology hubs. These are spaces where people can come together to learn, connect, and access the technologies they need to succeed. These hubs provide an inclusive and supportive environment, and they help to bridge the SocioEcTech divide, connecting people with the technologies they need to reach their full potential.

The rise of social entrepreneurship is another way that socioectech justice is being promoted. Social entrepreneurs are using technology to solve complex social problems, and they are building businesses that are inclusive, sustainable and just. By working with communities to create technology-based solutions, these entrepreneurs are helping to create a world where everyone has the tools they need to succeed.

Socioectech justice is essential for human growth and development. By ensuring that everyone has access to the technologies and resources they need to participate fully in society, we can build a world that is more inclusive, equitable and just. This chapter has explored the importance of socioectech justice, and how it can be promoted through community technology hubs, social entrepreneurship and other means. The future of inclusive socioeconomics depends on our ability to create a world where everyone has value-added tools they need to survive.

As we have seen in the cases of Mrs. Smith and the vulnerable individual who was falsely accused, the importance of socioectech justice cannot be overstated. It is crucial in ensuring that all individuals, regardless of their socio-economic status or technology literacy, are given equal opportunities and are not at risk of being marginalized or mistreated because of the digital divide. However, not everyone is as fortunate as Mrs. Smith, who received the help she needed. This highlights the urgency and necessity of addressing socioectech justice on a larger scale. And that is why it's time to delve deeper into the purpose and goals of this book, and explore ways in which we can work towards a more inclusive and just society.

6

Defining the Purpose and Goals of the Book

The purpose of this book is to shed light on the importance of socioectech justice and how it plays a crucial role in human growth and development. We have seen the stark contrast between the lives of Mrs. Smith and the vulnerable members of society who were falsely accused, and it is evident that the divide between those who have access to technology and those who do not can greatly impact one's life.

As a society, we have a responsibility to ensure that every person has access to the resources and opportunities they need to live a fulfilling life. This includes access to technology and the ability to participate in the digital world. The unequal distribution of technology creates a divide that can have serious consequences for individuals, families, and communities.

It is our goal to educate and raise awareness about the importance of socioectech justice and how it can have a positive impact on human growth and development. We aim to encourage those in positions of power to take action and make meaningful changes that will promote access, diversity, and inclusion in the technology sector.

We also want to inspire a movement for change and encourage individuals to advocate for socioectech justice in their communities. We believe that by working together, we can create a future where everyone has access to the tools and resources they need to thrive.

The principles of socioectech justice are rooted in the idea that every child needs a family, every family needs their day in court, and even the most vulnerable need access. This is especially true between the tales of two cities, where the divide between those who have access to technology and those who do not can be especially pronounced.

Meet little Johnny Smith, a six-year-old caught in the crossfire of a bitter custody battle between his parents, Sarah and Bob. Johnny's parents have been separated for a few years now, and their relationship has only become more hostile over time. They've been battling in court over custody of Johnny, and the process has become grueling and drawn-out, with each side hiring their own lawyers and fighting tooth and nail over every detail.

Johnny's parents are both represented by high-powered lawyers who are using every tool at their disposal to win the case. This includes using technology to gather evidence and make their case to the judge. Unfortunately, Johnny's parents are on different sides of the socioectech divide, with Sarah having access to cutting-edge technology and Bob struggling to keep up.

This divide has only heightened the adversity between Johnny's parents and their legal teams. Sarah's lawyers are using complex algorithms and artificial intelligence tools to sift through mountains of data and find evidence that will give her the upper hand in the case. Bob, on the other hand, doesn't have access to these tools and is struggling to make his case to the judge.

The situation is further complicated by the pandemic and the shift to virtual court proceedings. Sarah's lawyers are able to present their case in a highly organized and polished manner, leveraging their technology to make their case in a clear and concise way. Bob, on the other hand, is unable to attend the virtual hearing in person, as he doesn't have a computer or the necessary software to participate. This puts him at a significant disadvantage, as the judge is unable to see his expressions or hear his voice, leaving him at the mercy of Sarah's technologically savvy legal team.

As the custody case wears on, Johnny becomes more and more distressed by the constant arguing and fighting. He feels caught in the middle, with both parents fighting for him but also fighting against each other. The situation is made even worse when the clerk's office rules in Sarah's favor, accusing Bob of being a danger to Johnny based on the evidence presented by Sarah's lawyers.

During mandated e-filings due to the pandemic, circuit clerks who are on the inside of the court system have the power to accept, reject, and otherwise manipulate paper motions or paper answers to motions filed by individuals from the outside, who are working from home. This power dynamic can be concerning because the people on the inside of the court system have a certain level of technical knowledge and access to technology that those on the outside may not have. This can lead to situations where vulnerable members of society, such as the elderly or those without access to technology, may struggle to navigate the court system and have their voices heard. For example, if a person without access to a computer or the necessary software requirements is trying to file a motion or answer to a motion, their document may be rejected by the circuit clerk. This can lead to situations where the individual may be found guilty of a false accusation in their absence, as was the case with Mrs. Smith in chapter 4. It is important to address these issues of socioectech justice in order to ensure that everyone, regardless of their access to technology, has a fair and equal chance to navigate the court system and have their voices heard.

This is where the principles of socioectech justice come into play. The idea that every child needs a family, every family needs their day in court, and even the most vulnerable need access, is one that needs to be upheld in cases like Johnny's. The socioectech divide between Sarah and Bob only exacerbates the situation and leaves Bob without a voice in the virtual court proceedings. The principles of socioectech justice call for accessibility, diversity, and transparency in the use of technology, so that cases like Johnny's can be decided fairly and justly.

In Johnny's case, a group of concerned citizens came together to help bridge the socioectech divide. They met with Bob at the local library and taught him how to use the computer and software necessary to participate in the virtual court proceedings. This was a game-changer for Bob, as he was finally able to present his case in a way that was heard and understood by the judge. With the help of this peer group, Bob was able to demonstrate that he was a loving and responsible father who was capable of providing a safe and stable home for Johnny.

The case of Johnny Smith serves as a powerful reminder of why socioectech justice matters. It highlights the importance of ensuring that everyone, regardless of their socioectech status, has access to the tools and resources they need to participate fully in society. Without socioectech justice, the most vulnerable members of our society are at risk of being left behind and denied their day in court. By working together to promote socioectech justice, we can build a more inclusive and equitable society where everyone has the opportunity to succeed and access life decisions.

The principles of socioectech justice, including accessibility, diversity and inclusion, adaptability, transparency and accountability, are foundational to practices of socioectech justice, including promoting access to technology, fostering diversity in the technology sector, and advocating for ethical technology regulations.

It is our hope that this book will serve as a call to action for those who are passionate about promoting equity and justice in the technology sector. By working together, we can create a more inclusive and equitable future for all.

7

Examining the SocioEcTech Divide: Understanding the Digital Divide

The digital divide refers to the unequal distribution of technology and internet access between different groups of people. In the current technological age, access to the internet and technology has become essential for many aspects of daily life. However, not everyone has equal access to these resources, leading to a divide between those who have the resources and those who do not. This divide is known as the digital divide, and it can have a significant impact on human growth and development.

Meet Sarah, a single mother of two young children living in a low-income neighborhood. Sarah has always struggled to make ends meet, but the pandemic has made things even harder for her. Her children's school has gone virtual, and she needs to find a way to keep up with their education while also trying to make a living. However, Sarah doesn't have access to a computer or the internet, making it nearly impossible for her children to keep up with their school work.

Sarah has tried to find ways to access technology, but she can't afford it. She has tried to use the computers at the library, but they are often in high demand, and she can't always find an available one. She has even tried to use her phone, but the small screen and slow internet connection make it nearly impossible to complete school assignments or job applications.

Sarah's children are falling behind in their studies, and her job prospects are becoming more limited. She feels trapped and hopeless, unable to provide for her family and give them the resources they need to succeed. This is just one example of the impact the digital divide can have on human growth and development.

This divide can lead to limited access to education, job opportunities, and even healthcare. Those without access to technology are at a disadvantage, unable to participate in the digital economy and unable to take advantage of the many resources available online. This not only limits their personal growth and development but can also have a significant impact on their communities, perpetuating the cycle of poverty and inequality.

Take the case of two young students, Jessica and Jack, who just graduated from high school. They both apply for jobs at Caterpillar, a global manufacturing company. The application process is completely online and requires 152 steps, from uploading a resume to completing various assessments and tests.

Applying for a job can be a complicated process, with various steps to complete. One must create an account and log in, complete a personal information form with name, address, and contact information, upload a resume and cover letter, answer pre-screening questions about work history, education, and qualifications, complete a personality test or psychological evaluation, provide references and their contact information, agree to the terms and conditions of employment, and finally submit the application. After submitting, one must wait for a response from the employer and check the status of the application online through multiple sources.

Jessica, who has a computer and internet access at home, successfully completes all the steps and submits her application. However, Jack, who does not have access to a computer or internet, is unable to complete the process. He misses out on the opportunity to apply for the job and further his career. This is a sad example of how the digital divide can impact people's lives and their chances for success. The divide between those who have access to technology and those who do not is significant, and it can limit opportunities for personal and professional growth.

Jessica had the required technical, computer, software training, and internet training major contributors to digital bridging the divide, which can impact an individual's ability to access opportunities and participate in society. These barriers, training or not, often include a lack of access to reliable and affordable technology such as computers and high-speed internet, inadequate training in using these technologies, and limited knowledge of computer skills. Additionally, these barriers also encompass the technical limitations of hardware, software, and websites, which can make navigating the digital world challenging and frustrating for those without proper training and support. These obstacles not only limit access to job opportunities and education but also hinder civic engagement, community participation, and overall human growth and development.

The digital divide is a critical issue that must be addressed to ensure inclusive socioeconomics and promote human growth and development. socioectech justice must prioritize bridging this divide and ensuring that everyone has equal access to technology and the resources it provides.

The digital divide has been identified as a major challenge for inclusive socioeconomic growth. The internet has become a platform for information, communication, and commerce, making it a crucial resource for individuals, businesses, and communities. Without access to the internet, individuals are limited in their ability to connect with others, obtain information, and participate in economic activities. This can have a negative impact on their personal growth and development, as well as their ability to contribute to society.

The digital divide can be seen in many different ways. For example, urban areas typically have higher levels of internet access and technology resources than rural areas. Low-income individuals and communities also tend to have lower levels of internet access compared to higher-income individuals and communities. Additionally, elderly individuals and those with disabilities may have difficulty accessing technology due to physical or cognitive limitations.

The digital divide can also impact communities in different ways. For example, communities that lack access to technology may not be able to participate in online education or training programs, limiting their opportunities for personal and professional development. This can lead to a cycle of poverty and inequality, as individuals who lack access to technology and the internet may be unable to obtain jobs or participate in the economy.

In addition to the digital divide, there is also a divide in the quality of internet access. Individuals and communities who have access to the internet may still face issues such as slow connection speeds, limited data usage, or poor infrastructure. This can limit their ability to participate in online activities and access important resources, such as healthcare information or job opportunities.

The digital divide has far-reaching consequences for human growth and development, and it is crucial to understand and address these consequences in order to promote inclusive socioeconomic growth. The purpose of this book is to explore the impact of the digital divide, as well as the principles and practices of socioectech justice, in order to provide a roadmap for creating a more inclusive and equitable society.

The next chapter will examine the gap between the "haves" and "have-nots" and its impact on human growth and development. It is important to understand the root causes of the digital divide and its effects in order to develop effective strategies for promoting socioectech justice. The goal of this book is to provide a comprehensive understanding of the digital divide and the role of socioectech justice in promoting inclusive socioeconomic growth and development.

8

The Widening Gap between the Haves and Have-Nots

The digital divide is a major concern in today's society as it continues to widen the gap between the haves and have-nots. With technology playing an increasingly important role in our daily lives, those who lack access to technology are at a significant disadvantage. This divide not only affects individual opportunities for employment and education, but it also has broader implications for society as a whole.

The digital divide is characterized by a difference in access to digital technologies and the skills needed to use them. While some people have access to high-speed internet, computers, and the latest software, others are limited to outdated technology or even have no access to technology at all. This divide is particularly pronounced in lower income communities where families cannot afford the latest technologies and may lack the resources to purchase a computer or pay for internet access.

The social differences of the haves and have-nots

Meet Mrs. Jones, a single mother of two who lives in a low-income neighborhood in the inner city. Despite having a high school diploma and working two jobs, she struggles to make ends meet and provide for her children. She does not have access to a computer or the internet, and she relies on the local library for online resources.

On the other hand, there is Mr. Smith, a successful businessman who lives in a gated community in the suburbs. He has a PhD, a high-paying job, and a large house with all the latest technology. He has a personal computer, a laptop, and a tablet, and he is always connected to the internet.

Despite their vastly different backgrounds and circumstances, both Mrs. Jones and Mr. Smith are searching for a better life for themselves and their families. But the digital divide between the haves and have-nots presents a significant barrier to achieving their goals.

For Mrs. Jones, the lack of access to technology and the internet makes it difficult to apply for jobs, complete online applications, and stay connected with her children's school. She is at a disadvantage compared to those with access to these resources, and she is more likely to be left behind in the job market.

On the other hand, Mr. Smith's access to technology and the internet provides him with endless opportunities and resources. He can work from home, connect with colleagues from around the world, and access a wealth of information and resources. He is also better equipped to navigate the online job market, which is increasingly becoming the norm.

The social differences between the haves and have-nots are a testament to the widening gap between those who have access to technology and those who do not. This divide has far-reaching implications for human growth and development, and it is essential that we find ways to bridge the gap and provide equal access to technology and the internet for all.

The economic differences of the haves and have-nots

Mrs. Johnson is a single mother of two living in a low-income neighborhood in the city. She struggles to make ends meet and often goes without food so that her children can eat. Despite working long hours at a minimum wage job, Mrs. Johnson can barely afford the rent on her apartment and often falls behind on utility bills.

On the other hand, Mr. Smith is a successful businessman who lives in an upscale neighborhood in the suburbs. He drives a luxury car, wears expensive clothes, and enjoys a lavish lifestyle. Mr. Smith has a high paying job and a large savings account, allowing him to live a life of comfort and security.

The economic divide between Mrs. Johnson and Mr. Smith is a stark example of the widening gap between the haves and have-nots. Mrs. Johnson's financial struggles are not unique, as many people in low-income neighborhoods struggle to make ends meet. On the other hand, people like Mr. Smith are enjoying the benefits of a thriving economy, reaping the rewards of their hard work and investments.

This divide can have serious consequences for human growth and development, as people who live in poverty are more likely to suffer from health problems, face barriers to education, and experience limited access to opportunities and resources. The widening gap between the haves and have-nots creates an unequal society, where some people are given an unfair advantage while others are left behind.

This case study highlights the need for inclusive socioeconomics, where everyone has access to opportunities and resources, regardless of their financial status. By creating a more equitable society, we can help to reduce the divide between the haves and have-nots, promoting human growth and development for all.

Meet Mr. and Mrs. Smith, a couple going through a divorce. Mr. Smith is a highly successful businessman who has always had access to the latest technology and resources. He is tech savvy and has a team of lawyers to represent him in court. On the other hand, Mrs. Smith is a homemaker who has limited access to technology and resources. She is not tech savvy and has limited legal representation in court.

During their divorce proceedings, Mr. Smith used his resources to file all of his motions and answers to motions electronically through the court's e-filing system. Mrs. Smith, on the other hand, was unable to file her responses electronically because she did not have access to a computer or the necessary software. She was forced to rely on a legal aid clinic to help her file her responses, which resulted in her being at a disadvantage in court.

The court ultimately issued a decision that favored Mr. Smith. This decision was based on the fact that all of Mr. Smith's responses to motions were filed electronically and were easily accessible to the judge. Mrs. Smith's responses, on the other hand, were filed on paper and were not as easily accessible to the judge.

This case highlights the justice differences between the haves and have-nots. In this case, Mr. Smith had access to the latest technology and resources, which gave him a significant advantage in court. Mrs. Smith, on the other hand, was at a disadvantage because she did not have access to the same resources. This case demonstrates the importance of socioectech justice and the need for equal access to technology and resources for all individuals, regardless of their social and economic status.

Mrs. Smith and Mr. Smith were once a happily married couple. They had a daughter named Olivia who they both loved and cherished. However, their marriage took a turn for the worse and they eventually decided to divorce. The divorce process was long and arduous and was characterized by a lot of animosity between the two parties.

During the divorce proceedings, Mrs. Smith found herself at a disadvantage. She was not as well-educated as Mr. Smith and did not have the same access to technology and resources. She also had limited financial resources and was struggling to make ends meet. On the other hand, Mr. Smith was socially and economically advantaged. He had a good education, access to technology and resources, and a well-paying job.

The court proceedings were not fair to Mrs. Smith either. The court granted Mr. Smith full custody of their daughter, Olivia, and labeled Mrs. Smith as the noncustodial parent. This meant that she would only have limited access to her daughter and would have to pay child support. The court's decision was based on the fact that Mr. Smith was seen as being in a better position to provide for their daughter.

Mrs. Smith was devastated by the court's decision. She felt like she was being punished for not having the same advantages as Mr. Smith. She was also upset that she would only have limited access to her daughter and that her relationship with her daughter would be strained.

The situation only got worse for Mrs. Smith as the years went by. Mr. Smith's advantages only grew, while Mrs. Smith's disadvantages persisted. This meant that Mrs. Smith was unable to keep up with the child support payments and found herself in a difficult financial situation. The court order also meant that she was unable to see her daughter as often as she would like and that her relationship with her daughter was strained.

This case highlights the widening gap between the haves and have-nots in society. The differences in education, access to technology and resources, and financial situation between Mrs. Smith and Mr. Smith had a significant impact on the outcome of their divorce proceedings. The court's decision to grant Mr. Smith full custody of their daughter was based on these differences and had a lasting impact on Mrs. Smith's life. It also highlights the need for socioectech justice, as the most vulnerable members of society are often at a disadvantage in the justice system.

Mrs. Smith and Mr. Smith were once a happily married couple with a daughter named Olivia. However, their marriage began to deteriorate and they eventually decided to divorce. The divorce proceedings were a contentious affair, with both parties fighting for custody of their daughter. Mrs. Smith, who was a homemaker, had limited resources compared to Mr. Smith who was a well-educated professional. Mrs. Smith also had limited access to technology and was not familiar with the online divorce process, whereas Mr. Smith had access to all the latest technology and was proficient in using it.

The divorce proceedings were complicated by the COVID-19 pandemic, which mandated virtual court appearances and online filings. Mrs. Smith, who did not have access to a computer or the internet, struggled to participate in the virtual proceedings. On the other hand, Mr. Smith, who was tech-savvy, was able to use his resources to present his case effectively.

In the end, the court ruled in favor of Mr. Smith and he was granted full custody of their daughter, with Mrs. Smith being granted limited visitations. Mrs. Smith was devastated by the decision and felt that she had been unfairly treated by the court.

Years passed and Mrs. Smith passed away, but the court order remained in place. Mr. Smith continued to have full custody of their daughter, who grew up to be a successful professional. However, Mrs. Smith's family never forgot about her and felt that she had been unjustly treated in the divorce proceedings.

The case of Mrs. Smith and Mr. Smith highlights the vast differences between the haves and have-nots in our society, particularly when it comes to justice. Mrs. Smith's lack of access to technology and her limited resources put her at a disadvantage in the divorce proceedings. On the other hand, Mr. Smith's access to technology and his well-educated background allowed him to present his case effectively and ultimately win custody of their daughter. This case serves as a stark reminder of the importance of socioectech justice in ensuring that all individuals, regardless of their background, have access to the resources they need to participate in the justice system.

Moreover, the digital divide can also be seen in the skills and knowledge necessary to use digital technologies. Those who do not have access to technology early in life are often not exposed to the basic computer skills that are so important in the job market. Additionally, those who do not have access to technology are not able to take advantage of online resources for education and training, which can greatly limit their career and educational opportunities.

The effects of the digital divide are far-reaching and long-lasting. Those who are left behind are more likely to experience poverty, unemployment, and reduced economic mobility. The divide also exacerbates existing social and economic inequalities and creates a society that is less inclusive and more divided.

In conclusion, the digital divide is a critical issue that requires immediate attention. Closing the gap between the haves and have-nots will require a concerted effort from government, businesses, and community organizations. Access to technology must be seen as a fundamental right, and we must ensure that everyone has access to the tools and resources needed to participate fully in the digital age. Only then can we create a society that is truly inclusive and equitable, where everyone has the opportunity to reach their full potential and contribute to human growth and development.

As the divide between the haves and have-nots continues to widen, it becomes increasingly important to understand the negative consequences of inequality on human growth. The unequal distribution of resources, opportunities, and access to justice can have far-reaching impacts on individuals, families, and communities. In chapter 8, we will delve deeper into the negative consequences of inequality, exploring how it affects individuals and their ability to reach their full potential.

9

The Negative Consequences of Inequality on Human Growth

Inequality has long been a topic of discussion and concern for many individuals, organizations, and governments around the world. It has been widely recognized that inequality can have negative consequences for human growth and development. In this chapter, we will examine the ways in which inequality affects human growth and discuss the negative consequences of inequality on individuals and society as a whole.

One of the most obvious negative consequences of inequality is the economic impact it can have on individuals and communities. Individuals who are economically disadvantaged are more likely to experience poverty and lack access to basic necessities such as food, clothing, and shelter. This lack of access to basic necessities can lead to health problems, malnutrition, and other negative outcomes.

When talking about the negative consequences of inequality, the economic impact stands out as one of the most significant. Individuals who are struggling economically often face poverty, lack of access to basic necessities such as food, clothing, and shelter. This can lead to a vicious cycle of poverty, where it becomes more and more difficult to escape.

For example, a family living in poverty may not have enough money to purchase healthy food, leading to malnutrition and health problems. Children in these families may miss out on important nutrients and vitamins that are essential for their physical and mental development, putting them at a disadvantage compared to their peers. Furthermore, the lack of access to proper housing can result in overcrowding, poor living conditions, and exposure to health hazards.

Meet the Johnson family, a family of five living in a low-income neighborhood in the city. The parents, Mr. and Mrs. Johnson, both work minimum wage jobs and struggle to make ends meet each month. Despite their best efforts, they often have to choose between paying for food, rent, or utilities. This means that the family often goes without proper nutrition and may only have access to unhealthy, processed foods that are cheap and readily available.

The impact of this lack of access to nutritious food is evident in their children, who frequently suffer from malnutrition and health problems. The oldest child, Jessica, has struggled with stunted growth and lacks energy due to a lack of essential vitamins and nutrients. The middle child, James, has a history of respiratory infections and other health problems due to the poor living conditions in their overcrowded apartment.

This cycle of poverty and lack of access to basic necessities has a significant impact on the Johnson family, who are constantly trying to overcome the negative consequences of inequality. It is a clear example of how inequality can have a lasting impact on human growth and development, and the importance of addressing these issues through the principles of socioecotech justice.

In addition to the health impacts, economic disadvantage can also lead to reduced access to education and job opportunities. Children from low-income families are less likely to receive a quality education, making it difficult for them to break the cycle of poverty. Adults who are struggling economically may not be able to afford the necessary training or education to secure a better paying job, limiting their career opportunities.

Juan is a 17-year-old student who has grown up in a low-income family. He has always struggled with basic necessities, such as food and clothing, but has always dreamed of getting a quality education. Despite his hard work and dedication in school, Juan's family could not afford to send him to a private school or pay for extracurricular activities.

Juan attended a public school that was underfunded, lacking resources, and had overworked teachers. This limited his access to technology, textbooks, and adequate facilities. Furthermore, he had to work part-time to help support his family, taking away time from his studies and hindering his academic performance.

Despite his best efforts, Juan struggled to keep up with his peers, who had access to better resources and a more stable home environment. He fell behind in his classes and began to lose confidence in himself. By the time he graduated, Juan had missed out on important opportunities and did not have the necessary skills and education to secure a good job.

Juan's story is a common one, and highlights the negative consequences of inequality on human growth and development. Lack of access to quality education, resources, and job opportunities perpetuates the cycle of poverty, limiting the potential of individuals and communities.

Overall, the economic impact of inequality can be devastating, affecting not only individuals but entire communities. The lack of access to basic necessities, quality education, and job opportunities can create a cycle of poverty that is difficult to escape, leading to long-term negative consequences on human growth and development.

Furthermore, inequality can also affect the educational opportunities available to individuals. Those who are economically disadvantaged may not have access to quality education and opportunities to pursue higher education, which can limit their career prospects and earning potential. This lack of access to education can also perpetuate the cycle of poverty and limit the economic opportunities available to future generations.

Another way that inequality affects human growth is through the impact it has on social and political systems. Inequality can lead to social and political instability, as individuals and communities who feel that they are being treated unfairly may engage in protests or other forms of civil unrest. This instability can have a detrimental effect on the overall health and well-being of a society.

In the case of John and Maria, they come from two different socioeconomic backgrounds. John was born into a wealthy family and had access to the best schools, technology, and resources. On the other hand, Maria was born into a low-income family and had limited access to quality education and resources. After they got married and had a daughter, they got a divorce and the custody battle was intense. John, being economically advantaged, was able to hire the best lawyers and argue his case effectively in court, resulting in him being granted custody of their daughter.

This decision had a profound impact on Maria's life. She was devastated at the loss of her daughter and felt that the justice system was unjust and unfair to her. She felt that her socioeconomic background had played a significant role in the outcome of the custody battle. This feeling of unfairness and frustration fueled her desire to become involved in social activism, to fight against the systemic inequalities in society.

The case of John and Maria highlights the negative consequences of inequality on human growth. The inequality in the justice system, combined with the economic divide, can lead to mounting social and political instability. This instability can have a detrimental effect on the overall health and well-being of a society, as individuals and communities who feel that they are being treated unfairly may engage in protests or other forms of civil unrest.

In addition, inequality can also lead to a decline in public health. Individuals who are economically disadvantaged may not have access to quality healthcare, which can lead to an increase in health problems and diseases. This can result in a decrease in overall life expectancy, which has significant implications for human growth and development.

A case study of the negative consequences of inequality on public health can be seen in the disparities between healthcare access in urban and rural areas. In rural areas, there may be limited access to healthcare facilities and providers, making it difficult for residents to receive adequate medical care. This can result in a higher incidence of preventable diseases and conditions, such as heart disease, diabetes, and cancer.

Additionally, the lack of access to preventative care and regular check-ups can lead to more serious health problems that go undiagnosed and untreated. This can ultimately result in a decrease in overall life expectancy for residents in rural areas, and a decline in public health for the community as a whole.

In conclusion, inequality has a significant impact on human growth and development. It can limit access to basic necessities, education, and healthcare, and can lead to social and political instability. The negative consequences of inequality on human growth and development must be addressed in order to ensure a better future for all individuals and communities.

As we have seen, inequality has far-reaching consequences on human growth and development, affecting everything from health and education to social and political stability. However, it is not just a problem faced by individuals and communities, but rather a global issue that affects us all. In the next chapter, we will explore the concept of socioectech justice, a term that encompasses the commonality of the challenges faced by individuals and communities around the world and the importance of addressing these inequalities. With the increasing impact of technology on our lives, the need for socioectech justice has never been more pressing. As we turn to chapter 9, let us delve into the idea of how we can come together to create a more just and equitable world.

10

The Commonality of SocioEcTech Justice

In the modern era, it is impossible to ignore the impact of technology on society. As the world becomes more connected, it is essential that we understand the ways in which technology can either exacerbate or alleviate the inequalities that exist within our communities. This is where the concept of socioectech justice comes into play.

Socioectech justice refers to the fair and equitable distribution of the benefits and drawbacks of technology among all members of society. It is a complex and multifaceted issue, but at its core, it is about ensuring that everyone has access to the tools and resources they need to succeed in a rapidly changing world.

The commonality of socioectech justice is that it affects everyone. Whether you are a wealthy CEO or a single parent struggling to make ends meet, the impact of technology on your life is undeniable. For those who have access to the latest technology, it can be a source of power and prosperity. For those who are left behind, however, it can be a source of frustration and inequality.

The good news is that we can do something about this. By understanding the importance of socioectech justice, we can work together to create a more equitable and inclusive society. This means ensuring that everyone has access to the same opportunities, regardless of their race, gender, income, or any other factor.

In chapter 9, we will explore the various ways in which socioectech justice plays a role in our lives. We will look at the impact of technology on different aspects of society, such as education, healthcare, and the workplace. We will also examine the ways in which governments and organizations can take action to promote socioectech justice, such as through policies, programs, and initiatives aimed at closing the digital divide.

By the end of this chapter, you will have a better understanding of what socioectech justice is, why it matters, and what you can do to help promote it. So buckle up and get ready for a wild ride through the exciting world of socioectech justice!

Once, there was a young boy named Max who lived in a modern world filled with technology and advancements. Max was living a normal life until one day, he was kidnapped by a man who had a personal vendetta against technology and the impact it had on society. The kidnapper kept Max in a remote cabin with limited access to technology and the outside world.

Max was raised in a world devoid of technology and he was never allowed to experience the wonders of the modern era. He grew up without knowing how to use a computer, cell phone, or even the internet. The kidnapper wanted Max to experience life without technology and see the impact it had on him.

Years went by and Max was finally rescued by the police. He was overjoyed to be reunited with his family and to experience the world once again. However, when Max was released, he quickly realized that his lack of exposure to technology and the modern world had a profound impact on his life. He struggled to keep up with the rapid pace of technology and was unable to find a job due to his limited skills.

Max's family tried to help him catch up by enrolling him in technology classes and providing him with the latest devices. However, it was a slow process, and Max was quickly falling behind. He felt like he was stuck in a world that he could not understand and was unable to participate in.

As time went by, Max's frustration grew, and he began to understand the kidnapper's motivations. Max realized that the lack of access to technology and the modern world had a significant impact on his life and the opportunities he had available to him. He became a strong advocate for technology justice and the importance of ensuring that everyone had access to technology, regardless of their background or socioeconomic status.

Max's story serves as a reminder of the critical role that technology plays in our lives and the negative impact that can result from being denied access to it. It highlights the importance of socioectech justice and the need for everyone to have equal access to technology and its benefits. Max's journey was a difficult one, but in the end, it led him to a greater understanding of the world and the importance of technology in shaping our lives.

There once was a wealthy CEO named Sarah, who had it all. She was born into a wealthy family, had the best education and was successful in her business. She learned about technology and how to use it in business ten years ago, and she thought she had a great advantage over others in the field. However, she was about to be surprised by a single parent named Maria.

Maria was a single mother who had a hard time making ends meet. She had a job as a waitress, but the salary was not enough to provide for her family. Despite her busy schedule, she found time to learn about new technology on her own at home. She spent hours reading articles, watching videos, and taking online courses. In just a year, she had become proficient in several tech skills, including website development, social media marketing, and data analysis.

With her new knowledge, Maria started a small online business selling handmade crafts. She was able to connect with customers from all over the world and her business quickly grew. In just a few years, Maria had built a successful e-commerce business and was even able to hire other people to help her run the company.

Sarah and Maria's paths crossed when their businesses started to compete for the same customers. Sarah was shocked to see how far Maria had come and how successful her business was. Sarah's company was still doing well, but she was no longer the leader in the field. Maria's company had surpassed Sarah's company in sales and customer satisfaction, thanks to Maria's innovative use of technology and her ability to connect with her customers.

This case study shows that success in business is not just about being born into wealth or having the best education. It is about hard work, determination, and a willingness to learn and adapt to new technology. Maria's story is an inspiration to those who are struggling to make ends meet and are looking for a way to improve their lives. With the right mindset and determination, anyone can achieve success and make a better life for themselves and their families.

With the rise of technology and its impact on the job market, it is becoming increasingly important to have a strong understanding and skillset in technology. Jane and Jack are two individuals who exemplified this, each in their own unique way.

Meet Jane and Jack, two individuals with vastly different backgrounds and experiences. Jane grew up in a wealthy family and had access to all the resources and opportunities she needed to succeed. She went to an Ivy League university, had a successful career, and was eventually appointed CEO of a large tech company. Jack, on the other hand, was raised by a single mother who struggled to make ends meet. Despite his challenges, Jack was driven and determined to succeed. He taught himself how to code in his spare time and eventually started his own tech company from his small apartment.

Despite their different beginnings, Jane and Jack had one thing in common: their commitment to creating a more equitable and inclusive society. They both recognized that everyone, regardless of their background, should have the opportunity to succeed. To this end, Jane used her position of power and influence to push for equal pay, benefits, and promotions for all employees, regardless of race, gender, or income. Jack, meanwhile, made sure that his company was accessible to people from all walks of life, offering flexible work hours and remote work options to make it easier for those who needed it.

In the end, both Jane and Jack's efforts paid off. Jane's company became a leader in promoting diversity and inclusion, while Jack's company grew at an exponential rate, eventually surpassing Jane's company in size and success. Through their efforts, Jane and Jack showed that when we work together to create a more equitable and inclusive society, everyone benefits.

In conclusion, chapter 9 has highlighted the significance of socioectech justice and the common thread that runs through all its aspects. We have seen how it encompasses not only technology and economics, but also social and political justice, and how it can be a force for good in the world. With this knowledge, we are now ready to delve deeper into one of the key aspects of socioectech justice: accessibility. In chapter 10, we will explore the importance of ensuring that everyone has access to the same opportunities and resources, regardless of their background or circumstances. So, let's continue this journey and see how we can work together to create a more just and inclusive society.

11

The Importance of Accessibility in SocioEcTech Justice

Access to technology has become a vital component of our daily lives. It influences how we communicate, how we learn, how we work, and how we grow. However, not everyone has equal access to these technologies and the opportunities they bring. This disparity in access is a critical issue that must be addressed if we hope to build a more just and equitable society. Chapter 10 explores the importance of accessibility in socioectech justice, highlighting the role it plays in creating a more inclusive world. The thesis of this chapter is that ensuring accessibility to technology and opportunities is fundamental to promoting socioectech justice and fostering human growth.

In today's world, technology is an integral part of our daily lives, and it has become a necessity in almost every aspect of our existence. However, despite the numerous benefits that technology has brought to society, many individuals are still being left behind due to barriers to entry. One of the biggest barriers to entry is accessibility, which refers to the ability of individuals to access, use, and benefit from technology. In order to create a truly inclusive society, it is essential to ensure that technology is accessible to everyone, regardless of their abilities or disabilities.

One of the groups that face significant barriers to entry when it comes to technology are individuals with disabilities. These individuals may have difficulty accessing and using technology due to physical, cognitive, or sensory limitations. For example, individuals who are blind or have low vision may have difficulty accessing websites and digital content that are not designed with accessibility in mind. Similarly, individuals with hearing or speech disabilities may struggle to use voice-activated technologies such as Amazon's Alexa or Google Home.

The lack of accessibility in technology not only limits the opportunities of individuals with disabilities, but it also perpetuates the cycle of inequality in society. In many cases, individuals with disabilities are unable to participate in the workforce or access educational opportunities, which can lead to a lower standard of living and a limited future. Moreover, the lack of accessibility in technology also affects individuals who are aging or who have temporary disabilities, such as a broken arm or a temporary illness.

Meet Sarah, a young girl with a hearing disability who was a part of a court case during the pandemic. Sarah's case was one of the many that were being heard remotely due to the restrictions put in place due to the pandemic. However, Sarah soon realized that the technology being used was not accessible to people with disabilities like her. She struggled to understand the proceedings and was unable to effectively communicate her side of the story to the judge and jury.

Despite her best efforts, Sarah lost her case as a result of the inaccessible technology. This left her feeling frustrated and helpless, as she felt that the technology had failed her in a crucial moment. She realized that the technology being used was not accessible to people with disabilities like her, and that this was a major barrier to entry that was preventing her from participating fully in society.

Sarah's story is a powerful reminder of the importance of accessibility in socioectech justice. It highlights the need for technology to be inclusive and accessible to everyone, regardless of their abilities or disabilities. This means taking into account the needs of people with disabilities and ensuring that the technology is designed and implemented in a way that is inclusive and accessible to everyone.

Sarah's study on the importance of accessibility in socioectech justice can be considered a "Marque Facility" in revitalizing inclusive socioeconomics for good. By exploring the barriers to entry for individuals with disabilities, Sarah sheds light on the need for increased accessibility in technology and the justice system. Her findings highlight the need for equal opportunities and support for all, regardless of abilities or disabilities, to ensure that everyone has a fair shot at success and well-being. This study is a valuable contribution to the efforts to create a more equitable and inclusive society, and serves as a call to action for those in the tech and justice industries to make the necessary changes to ensure accessibility for all.

To overcome these barriers to entry, it is essential to incorporate accessibility into the design and development of technology from the very beginning. This means ensuring that technology is accessible to everyone, regardless of their abilities or disabilities, and that it is designed with the needs of all users in mind. For example, websites and digital content can be designed with accessibility in mind by using clear and simple language, providing audio descriptions, and using high-contrast text and images. Similarly, voice-activated technologies can be designed with accessibility in mind by incorporating features that allow individuals with speech disabilities to use the technology.

When accessibility fails, it can have a detrimental impact on individuals and communities. This is particularly true in today's society where technology plays such a significant role in our daily lives. For people with disabilities, barriers to entry can be particularly pronounced, making it difficult for them to participate fully in society. When access is limited, it can lead to further marginalization and a lack of opportunities for growth and development.

However, with enough awareness and support, adaptability can play a crucial role in revitalizing inclusive socioeconomics. This involves creating flexible and responsive organic networks that are designed to accommodate a wide range of abilities and needs. By providing accessible technology, like regenerative (as opposed to unregenerative) resources, and support, people with disabilities can be empowered to participate fully in society and reach their full potential.

Chapter 12 of the book will delve deeper into this topic, exploring the importance of adaptability in creating a more inclusive society. Through case studies and real-world examples, readers will gain a better understanding of how adaptive technologies and support systems can play a key role in promoting equality and revitalizing inclusive socioeconomics.

Imagine a world where no one ever adapted beyond barriers, where everyone simply accepted their limitations and never attempted to overcome them. In this world, people with disabilities would be completely isolated and unable to participate in society. They would have limited access to education, employment, and healthcare, leading to a lack of opportunities for growth and development. This would result in a society where people with disabilities are marginalized and excluded from the mainstream, leading to a decrease in overall human growth and well-being.

However, in reality, people with disabilities have shown remarkable resilience and determination in the face of obstacles. They have found ways to adapt and overcome barriers, breaking down the limitations that have been imposed on them. This is what makes our world truly remarkable, and it is through these efforts that we can create a more inclusive, equitable, and accessible society for all.

Meet Sarah, a young woman born with cerebral palsy. Despite her physical limitations, Sarah was determined to succeed and make a difference in the world. Growing up, she faced numerous barriers to entry, including limited access to education and employment opportunities. However, Sarah refused to be held back by her disabilities and instead, used them to her advantage.

Sarah's passion for technology led her to study computer science in college, where she learned about the importance of accessibility and the ways in which technology could help people with disabilities overcome socioeconomics barriers. She then went on to start a tech company that specializes in creating accessible software and hardware for people with disabilities.

Sarah's company quickly gained recognition, and soon she was working with government agencies and large corporations to improve accessibility in their products and services. Her innovative solutions have helped countless people with disabilities lead more fulfilling lives, and she continues to work tirelessly to break down socioeconomics barriers for those with disabilities.

Through her determination and hard work, Sarah has become a marque facility in revitalizing inclusive socioeconomics and breaking down socioeconomics barriers for people with disabilities. By creating accessible technology, she is making it possible for people with disabilities to participate fully in society and to grow and develop to their full potential.

The importance of accessibility in socioectech justice cannot be overstated. By ensuring that technology is accessible to everyone, regardless of their abilities or disabilities, we can help to create a more inclusive and equitable society. This will help to break down the barriers to entry and provide equal opportunities for all individuals, allowing everyone to participate in the workforce and access educational opportunities. In the next chapter, we will delve deeper into the concept of accessibility in socioectech justice and explore some of the innovative solutions that are being developed to make technology more accessible to everyone.

12

Encouraging Diversity and Inclusion in Technology

The technology industry has come a long way over the past few decades, but there is still much work to be done in terms of promoting diversity and inclusion. In this chapter, we will explore the various ways in which technology companies can promote diversity and inclusivity in their workplace and product offerings.

One of the first steps in promoting diversity and inclusion is to make sure that your workplace is a safe and welcoming environment for all employees, regardless of their race, gender, sexual orientation, or any other factor. This means creating a culture of openness and respect, and providing resources and support to help employees feel comfortable and valued.

Let's explore the value of promoting diversity and inclusion in the tech industry. This is important because a diverse workforce and diverse perspectives in technology drive innovation and growth. With a more diverse group of people working in tech, we can expect better representation of different backgrounds, cultures, and perspectives. This leads to a more inclusive and fair society.

For example, imagine a tech company with a team made up of people from different countries, cultures, and backgrounds. These diverse perspectives can bring fresh ideas and unique solutions to problems, leading to breakthrough innovations and a more successful company. On the other hand, a homogeneous tech industry would limit the potential for new ideas and solutions, stifling growth and progress. By promoting diversity and inclusion in tech, we can create a more dynamic and thriving industry that benefits everyone.

To encourage diversity and inclusion in technology, let's focus on the crucial role of promoting diversity and inclusion in the tech industry. This includes everything from the creation of technology products to the hiring and promotion of employees. The reason why this is so important is because promoting diversity and inclusion leads to a more innovative and thriving tech sector.

For example, let's take the story of Sarah. Sarah is a young woman with a passion for technology and a drive to succeed. However, she faces barriers in the tech industry due to her background and lack of connections. Despite this, she is determined to succeed and finds a way to break through these barriers. With her hard work and determination, Sarah eventually becomes a successful software engineer at a large tech company.

Sarah's story highlights the importance of inclusive practices in technology. By promoting diversity and inclusion, we can ensure that everyone, regardless of their background or abilities, has the opportunity to succeed and contribute to the tech industry. This leads to a more inclusive and equitable society, where everyone has a chance to reach their full potential.

Another important step is to make sure that your products and services are accessible to everyone, regardless of their abilities or disabilities. This means using inclusive design principles and working closely with communities of people with disabilities to ensure that your products and services meet their needs.

Now the topic of diversity and inclusion in the tech industry. We understand the need for a diverse and inclusive work environment in the tech industry, not only for driving innovation but also for building a better society. To illustrate our point, let's consider the story of Sarah. Sarah was a young woman who was passionate about technology. She had a great idea for a start-up and wanted to enter the tech industry. However, she faced numerous barriers because of her gender and race. Despite her passion and drive, she was unable to secure funding or find a job in the tech industry.

This story highlights the challenges that the tech industry faces in promoting diversity and inclusion. The tech industry is often dominated by a homogenous group of individuals, making it difficult for people like Sarah to enter and succeed in the industry. In this book Revitalizing Inclusive Socioeconomics, we explore these challenges and discuss the importance of promoting diversity and inclusion in the tech industry. We also look at strategies for overcoming these challenges and highlight the positive impact that a more diverse and inclusive tech industry can have on the economy, society, and the world at large.

In addition to these steps, there are a number of other initiatives that technology companies can take to promote diversity and inclusion, such as supporting diversity organizations and initiatives, offering training and development programs, and mentoring programs for underrepresented groups. For example , Google's diversity and inclusion initiatives include the "Google for Startups" program, which provides support and resources to diverse founders and startups, and their "Diversity and Inclusion Fellowship Program," which provides mentorship and training to underrepresented groups in the tech industry. These initiatives not only help to promote diversity and inclusion in the tech industry, but also help to create a more inclusive and equitable society for everyone.

At the end of the day, promoting diversity and inclusion in technology is about more than just doing the right thing. It is about creating a more equitable and inclusive world for everyone, and helping to build a future in which technology can be used to benefit everyone, regardless of their background or abilities.

As we continue our journey towards revitalizing inclusive socioeconomics, it is important to acknowledge that despite our efforts to encourage diversity and inclusion in technology, there will still be challenges and barriers to overcome. This is why in chapter 12, we turn our focus to building adaptability through SocioEcTech justice. By understanding the importance of adaptability, we can ensure that marginalized communities and individuals with disabilities are equipped to overcome any obstacles and succeed in an increasingly digital world. Join us as we delve deeper into this crucial aspect of promoting inclusive socioeconomics.

13

Building Adaptability With SocioEcTech Justice

The world of technology is constantly changing, and with that change comes new challenges and opportunities. For society to thrive in this ever-evolving landscape, it is essential to build adaptability and resilience into our systems and processes. This is particularly important for those facing social, emotional, economic, or disability-related challenges, who are at risk of being left behind. In this chapter, we will explore the importance of socioectech justice in building adaptability and resilience in society.

One of the main challenges facing society today is the digital divide. This refers to the gap between those who have access to technology and those who do not. For people with disabilities, this divide can be particularly pronounced, making it difficult for them to participate fully in society. This can lead to further marginalization and a lack of opportunities for growth and development.

However, with the right support and resources, people with disabilities can break through these barriers. For example, assistive technology can help those with disabilities to better access and participate in society. Similarly, inclusive policies and practices in the tech industry can help to remove barriers for marginalized communities, such as those with disabilities or underrepresented minorities.

Lance was born with a physical disability that made it difficult for him to participate in activities that many people take for granted, such as playing sports, going to school, and finding employment. Despite the many challenges that he faced, Lance was determined to break through these barriers and make a meaningful impact on the world.

With the help of assistive technologies and the support of those around him, Lance was able to participate in a range of activities that he had previously thought were out of reach. For example, he was able to use a wheelchair-friendly sports program, which allowed him to play basketball, soccer, and other sports with other kids his age. He also attended a school that was equipped with the latest assistive technologies, such as voice recognition software and digital whiteboards, which helped him to participate in class and keep up with his studies.

In addition to the technological advancements, Lance also benefited from the social improvements that were taking place in his community. He was able to participate in a mentorship program that connected him with other kids who had disabilities, and he was able to connect with people who could provide him with emotional and practical support. With this support, Lance was able to grow into a confident young man who was able to pursue his dreams and make a difference in the world.

As Lance's story demonstrates, with the right support and resources, people with disabilities can break through the barriers of employment, education, and sports, and participate fully in society. By focusing on the socially, emotionally, economically, and disabilities challenges that must be broken through to advance as a more civilized society, we can help to create a world where everyone has access to the resources and opportunities that they need to thrive.

It is also important to consider the emotional and psychological impact of technology on society. As technology continues to change and advance, it is important to understand the impact this has on people's emotions, relationships, and mental health. This is particularly important for those facing social, emotional, or economic challenges, who may be more vulnerable to the negative effects of technology.

Jasmine was a bright and ambitious young woman who had always dreamed of pursuing a career in science and technology. However, her life took a dramatic turn when she was involved in a car accident that resulted in the death of both her mother and father. She was left with physical and emotional scars that made it difficult for her to move forward.

Despite these challenges, Jasmine refused to give up on her dreams. She knew that she had to adapt to overcome the barriers she was facing, and she found ways to do just that. For example, she discovered assistive technologies that helped her to better access and participate in society. She also found that the supportive community around her, along with at-home developments, helped her to remove some of the social and emotional barriers she was facing.

Jasmine also faced economic challenges, as she was left without a support system and had to rely on her own resources. However, she refused to let this stop her from pursuing her goals. She found ways to adapt, such as taking on a part-time job while she continued her studies. She also found support from mentorship programs and organizations that provided her with the resources and guidance she needed to overcome her challenges.

In the end, Jasmine's determination and resilience paid off. She was able to graduate with honors and secure a job in the technology industry, where she was able to make a real impact. Her story is a testament to the importance of adaptability in breaking through socio-economic barriers and revitalizing inclusive socioeconomics.

To build adaptability and resilience in society, it is essential to address these challenges and create inclusive systems that promote equity and justice for all. This requires a multi-faceted approach, including investment in research and development, public policies that promote access and equity, and a focus on the needs and experiences of marginalized communities.

At the Dallas, Texas Abilities Expo 2022, visitors were able to witness inspiring adaptive solutions. From adaptive child safety products created by United CLU parents for those who are systematically excluded, to adaptive rock climbing by an owner who became disabled from a rock climbing accident. Visitors also got to experience adaptive sports like wheelchair basketball, hockey, and water skiing, all demonstrated by individuals who were injured doing something they loved but still found a way to continue doing it and share their adaptations with others. Safety was a key factor in all the adaptive solutions, such as the adaptive water skiing booth that was accompanied by Albert Pruitt, who was in a wheelchair after a motorcycle accident, and the adaptive rock climbing booth that was accompanied by Mark Wiltman, who was in a wheelchair after a rock climbing accident. There were even adaptive transportation booths showcasing innovative solutions.

In conclusion, building adaptability and resilience with socioectech justice is essential for creating a more inclusive and equitable society. By addressing the challenges faced by marginalized communities and promoting inclusive practices in the tech industry, we can work towards creating a world where everyone has equal opportunities to succeed and grow.

As we've seen in Chapter 12, building adaptability through socioectech justice is a crucial step in ensuring that all members of society, including those with disabilities and facing socio-economic challenges, can fully participate and reach their potential. But as we move forward into the future, it's equally important that we have transparency and accountability in the technology sector. In Chapter 13, we'll delve into how fostering transparency and accountability in the tech industry can lead to more equitable and sustainable outcomes for all. So, buckle up and let's continue our journey towards a more inclusive and just future!

14

Fostering Transparency and Accountability in the Technology Sector

In recent years, technology has become an integral part of our daily lives. From social media platforms to e-commerce sites, technology has revolutionized the way we communicate, shop, and access information. However, as technology continues to advance, it is increasingly important to ensure that the technology sector is transparent and accountable in its practices. This chapter explores the ways in which the technology sector can foster transparency and accountability to promote trust, accountability, and ethical behavior.

One way in which the technology sector can promote transparency and accountability is through the adoption of open and transparent data policies. This means that companies should make their data available to the public and provide clear explanations of how they use and store data. This helps to promote trust and accountability among users and also helps to protect their privacy and security.

The Smith family

Mrs. Smith noticed that her son, Billy, was becoming increasingly distant and moody. After some investigation, she discovered that he was spending hours on his phone and computer, absorbing junk data from various sources. Concerned for her son's well-being, Mrs. Smith decided to take action. She sat down with Billy and had a heart-to-heart conversation about the impact of technology on his mental and emotional health. Together, they agreed to set limits on screen time and actively seek out positive and educational content for him to engage with. Mrs. Smith became a "data doctor" for her son, carefully curating the information that came into his devices and protecting him from negative influences.

The Johnson family

Mr. Johnson noticed that his daughter, Sarah, was becoming more and more distant and disengaged from her family. After some investigation, he discovered that she was spending hours on social media, comparing herself to others and feeling inadequate. Concerned for his daughter's well-being, Mr. Johnson decided to take action. He sat down with Sarah and had a conversation about the impact of social media on her self-esteem. Together, they agreed to set limits on screen time and actively seek out positive and uplifting content for her to engage with. Mr. Johnson became a "data doctor" for his daughter, carefully curating the information that came into her devices and protecting her from negative influences.

The Brown family

Mrs. Brown noticed that her kids were becoming increasingly addicted to their screens and detached from reality. Concerned for their well-being, she decided to take action. She sat down with her children and had a conversation about the impact of technology on their lives. Together, they agreed to set limits on screen time and actively seek out positive and educational content for them to engage with. Mrs. Brown became a "data doctor" for her family, carefully curating the information that came into their devices and protecting them from negative influences. She also taught her kids how to be mindful and present in the moment, and to value real-life experiences over virtual ones.

Another way to foster transparency and accountability in the technology sector is through the implementation of ethical business practices. This includes ensuring that technology companies have clear and transparent policies on data privacy, information security, and user protection. Companies should also have policies in place to address and prevent workplace harassment, discrimination, and other forms of unethical behavior.

One example of peer parents who organized to reverse the trend of lack of transparency and accountability in the technology sector is a group of parents in a suburban community in the United States. This group of parents came together after growing concerned about the amount of time their children were spending on their devices and the impact it was having on their mental and emotional well-being.

The group began to research the policies and practices of technology companies and were alarmed at the lack of transparency and accountability around data privacy and user protection. They decided to take action and started reaching out to other parents in their community to join them in their mission.

Together, they formed a parent-led organization with the goal of fostering transparency and accountability in the technology sector. The organization reached out to technology companies and began advocating for better privacy policies, more secure data management practices, and an overall commitment to ethical business practices.

The parent-led organization also began to educate other parents on the importance of these issues and how they could take action to make a difference. Through community events and educational seminars, the organization helped other parents to become more informed and empowered to make positive changes in the technology sector.

As a result of their efforts, the parent-led organization was able to bring about real change in the technology sector. Many of the technology companies they reached out to implemented new policies and practices to better protect user privacy and security. Additionally, the organization continued to grow, and other parents in other communities began to follow their lead, forming similar organizations and advocating for change.

In addition to these steps, the technology sector can also promote transparency and accountability by providing regular reports on the impact of their products and services on society. This can include regular reports on the impact of technology on employment, social and economic inequality, and the environment. This helps to promote greater accountability and ensures that the technology sector is contributing to the overall well-being of society.

A small town in rural America once relied heavily on a tech company that provided jobs and resources to the community. However, the company did not have clear policies on data privacy, information security, and user protection. Over time, several residents in the town began to experience issues with the company, from data breaches to workplace harassment and discrimination.

Despite the community's efforts to address these concerns and push for change, the company remained steadfast in their harmful practices. Eventually, the company went bankrupt, leaving the community in a state of crisis. The bankruptcy also had a ripple effect on the town, causing small businesses to shut down and leaving many residents without jobs.

Worst of all, the community was left with the aftermath of the company's unethical practices, from wrongful charges to the depletion of resources and even a decline in the overall quality of life. This tragedy serves as a cautionary tale of what can happen when a community fails to advocate for transparency and accountability in the technology policy-making sector.

Finally, the technology sector by itself fosters transparency and accountability by encouraging open and transparent communication with its users and stakeholders. This includes providing regular updates on company policies and practices and actively engaging with stakeholders to understand their concerns and feedback.

In conclusion, fostering transparency and accountability in the technology sector is crucial for promoting trust and ethical behavior. Through the adoption of open and transparent data policies, ethical business practices, regular reporting on the impact of technology, and open and transparent communication, the technology sector can contribute to a more just, equitable, and sustainable society.

As we have seen in chapter 13, fostering transparency and accountability in the technology sector is crucial for promoting ethical and responsible practices. However, simply having policies in place is not enough. In chapter 14, we delve deeper into the importance of implementing the principles of socioectech justice in order to drive real change and create a more inclusive and just society. It is time to move beyond simply talking about the problems and start taking concrete steps towards creating solutions. Let us now turn our attention to the practicalities of implementing socioectech justice in our communities and technology companies.

15

Implementing the Principles of SocioEcTech Justice

Welcome to chapter 14. If you've made it this far you're an amazing reader! Here we'll dive into the exciting and crucial topic of implementing the principles of socioectech justice! As we've learned throughout this book, the intersection of technology and society creates the need for a new framework of justice to ensure equal opportunities, protection, and representation for all.

So, what exactly does it mean to implement the principles of socioectech justice? Essentially, it means taking action to ensure that the benefits and impacts of technology are distributed equitably across all communities and individuals. This includes considering factors such as socioeconomic status, race, gender, ability, and more.

But, how do we go about implementing these principles in the real world? Well, the first step is to understand the unique challenges and barriers that different communities face when it comes to accessing and benefiting from technology. This could include issues like a lack of access to high-speed internet, inadequate training and education programs, or discriminatory hiring practices in the tech industry.

Once we understand these barriers, we can start to identify specific steps and initiatives that can help overcome them. This could include things like expanding broadband access to rural and low-income communities, providing training and development programs for underrepresented groups in the tech industry, or advocating for more inclusive hiring practices.

In the past, the implementation of socioectech justice principles was often lacking. For example, in the early days of the internet, many rural communities were left behind in terms of access and technology infrastructure. This created significant disparities in terms of access to information, resources, and opportunities.

In the present, there are many efforts underway to address these disparities and promote greater equity in access to technology. For example, many non-profit organizations are working to provide low-cost or free computer access to underserved communities. Additionally, some cities are implementing programs to provide free Wi-Fi in public spaces, making it easier for people without internet at home to access the internet.

In the future, it is important to continue these efforts and to ensure that technology is developed in a way that is inclusive and accessible for everyone. For example, the development of assistive technologies for individuals with disabilities will play a critical role in promoting greater equity in access to technology. Additionally, efforts to provide digital skills training and education will help to ensure that everyone has the skills and knowledge needed to participate fully in the digital economy.

It's important to note that implementing the principles of socioectech justice isn't just the responsibility of technology companies. We all have a role to play in ensuring that technology is used in an ethical and equitable manner. For example, we can support organizations and initiatives that promote diversity and inclusion in tech, advocate for privacy-protective policies, and make conscious decisions about the technologies we use and support.

Meet John, a tech CEO of a large tech company. John was solely focused on making profits and was not concerned with the ethical implications of the technology his company produced. His company developed a social media platform that harvested user data and sold it to advertisers without users' consent. He also failed to address workplace harassment and discrimination within his company, leading to a toxic work environment.

Despite receiving numerous complaints and media attention, John continued to prioritize profits over ethics and disregarded the consequences of his actions on both his employees and users. As a result, the company faced a major public relations crisis and eventually went bankrupt.

Meet Sarah again, a tech entrepreneur and CEO of a small start-up company. Sarah was dedicated to creating technology that positively impacted society and was mindful of the ethical implications of her actions. Her company developed an app that aimed to reduce food waste and support local farmers.

Sarah was passionate about promoting diversity and inclusion in tech and ensured that her company had clear policies in place to address and prevent workplace harassment, discrimination, and other forms of unethical behavior. She also made a conscious effort to use privacy-protective technologies and was transparent about her company's data collection practices.

Sarah's company was successful and had a positive impact on the environment and local communities. She was praised for her ethical leadership and her company served as a model for other tech companies to follow.

Another example of a hero using socioectech justice to right a wrong is Ella, a tech engineer who worked for a large tech company that had a history of discriminatory practices towards employees of color and women. Ella was determined to use her skills and knowledge to address these injustices. She began by organizing and leading a group of diverse employees within the company to advocate for policies and practices that would promote equity and diversity. This led to the creation of diversity and inclusion committees, and the implementation of policies to ensure fair hiring practices, equal pay, and professional development opportunities for all employees.

Ella was not content with just making changes within her own company, she also spoke out publicly about the need for the tech industry as a whole to take responsibility for creating a more inclusive and equitable environment. Her activism and leadership helped to bring attention to the issues of diversity and inclusion in the tech industry and inspired other tech companies to take similar steps towards creating a more just and equitable workplace.

Ella's dedication to the principles of socioectech justice and her tireless efforts to promote equity and justice within the tech industry earned her recognition as a hero in the field, and she continues to inspire others to work towards a more equitable future through the use of technology.

Ella's unwavering commitment to the principles of socioectech justice had inspired many people in her community. She believed that technology should be used for the greater good and that its benefits should be shared equitably among all individuals and communities. However, not everyone shared the same values and principles as Ella.

Enter Maxless, a successful entrepreneur who seemed to have it all. He was charismatic, charming, and driven. However, behind his glossy exterior was a man who was only interested in making money, no matter the cost. He was a master at exploiting others and making a fortune at the expense of those who trusted him. In the world of technology, he found a new opportunity to further his wealth by manipulating the system and misleading people. Maxless had no regard for the consequences of his actions, and he was completely blind to the principle of socioectech justice. He was the polar opposite of Ella and everything she stood for.

Once, there was a villain named Maxless. Maxless was a successful entrepreneur who saw a gap in the market for a new technology that promised to revolutionize the world. Maxless was ambitious and determined to make a fortune from his new invention, but he was willing to do whatever it took to achieve his goal, even if it meant exploiting others.

Maxless launched his new technology, which he called "SociEcTech Justice", and claimed that it would be a platform for fair and equitable distribution of technology benefits to all communities and individuals. He convinced many people, including investors and customers, to support his new venture by promising them a share of the profits and the benefits of his new technology.

However, Maxless was not as honest as he appeared. He had a hidden agenda, which was to control and manipulate the use of technology for his own selfish interests. He started collecting personal data and information from his users, without their knowledge or consent. He used this data to manipulate the market and influence the decisions of his users, to make a profit for himself.

Maxless also convinced his customers to invest their money in his company, promising them high returns on their investments. However, he was secretly diverting the funds to his own personal account, leaving his customers with nothing.

As time went by, Maxless became even more bold in his actions. He used his control over the technology to spread false information and propaganda, misleading people into making decisions that would benefit him. He also threatened his competition and anyone who dared to question his actions, by using the power of his technology.

Eventually, Maxless's actions caught up with him, and he was exposed as a fraud. His company was declared bankrupt, and he was sentenced to prison for his crimes. His customers were left with nothing, and many of them lost their life savings. The people who had trusted him were devastated and felt betrayed.

The story of Maxless serves as a cautionary tale, about the dangers of using technology for unethical and selfish purposes. It highlights the importance of using technology in an ethical and equitable manner, and of implementing the principles of socioectech justice. It reminds us that we all have a role to play in ensuring that technology is used for the greater good, and that we must be vigilant in holding those who abuse technology accountable.

As the story of Maxless highlights, the exploitation of individuals and communities through technology is not uncommon in today's society. However, this serves as a cautionary tale and reminder that the implementation of socioectech justice requires a collective effort and cannot be left to chance. For example, we have seen the devastating consequences of the painkiller epidemic where Big Pharma companies put profit over people, leading to widespread addiction and death. The family law system also perpetuates issues rather than solving them, causing harm to families and children. The power of technology has the potential to make a positive impact on the world, but without proper ethical considerations, it can also lead to negative consequences. Remember when we were young and had dreams of making the world a better place? We cannot let those dreams be taken away by those who seek to exploit the system for their own gain. It is up to us, as a collective, to advocate for access to technology for all and ensure that its benefits are shared equitably across all communities and individuals. The principles of socioectech justice must be at the forefront of this movement in order to create a better future for everyone.

This is where individuals like Ella come into play as heroes in the face of socioectech injustices. Ella's dedication to the principles of socioectech justice, combined with her passion for using technology for good, sets her apart as a leader in the fight for a more equitable world. Through her actions and advocacy, she serves as a shining example of what is possible when we come together to make positive change.

In contrast, an example of a law school graduate who went into advertised family law practice is Sarah. Sarah promised her clients that she would handle their divorce case with ease and that they would have a smooth process. However, after several court hearings, Sarah's clients began to realize that she was not as knowledgeable about the system as she claimed. She often failed to inform them of important deadlines, didn't keep them updated on the progress of their case, and frequently made mistakes in the court proceedings. This resulted in a lot of frustration and stress for her clients, and they ultimately lost faith in her ability to represent them.

On the other hand, another law school graduate, named Jack, went into advertised divorce representation with a different approach. Jack made sure to keep his clients informed of all the disconnects and fragmented thinking that came along with the divorce process. He was transparent about the challenges they would face and helped them understand the complexities of the system. Jack always made sure that his clients were up-to-date on the progress of their case, and he always fought for their rights in court. This honesty and transparency earned Jack a great reputation in the community and many clients would recommend him to others. Jack's approach of educating his clients about the process and keeping them informed resulted in a much smoother and less stressful experience for them.

Jack, the law school graduate who went into advertised family law practice, had a more open-minded and inclusive approach to his work. He was able to see the bigger picture and understand that every case was unique and required a different approach. He was able to connect with his clients on a personal level and understand their individual needs and concerns. Jack was able to collaborate with other professionals in the field, including judges, mediators, and other lawyers, to come up with the best solution for his clients.

On the other hand, Sarah, the fellow graduate who went into advertised "family law," seemed to live in a bubble. She was more isolated and seemed to only focus on her own goals and interests. She did not seem to be as empathetic as Jack, and did not connect with her clients in the same way. Sarah seemed to be more interested in winning the case rather than finding a solution that was in the best interest of her clients. She often cried wolf and overstated the issues in the case, which sometimes resulted in additional stress and frustration for her clients.

Overall, Jack's open-minded and inclusive approach to his work allowed him to provide better support and guidance to his clients. In contrast, Sarah's isolated and self-focused approach resulted in a less fulfilling experience for her clients and a less successful outcome for her practice.

In order to make a meaningful impact, we must all take a stand and advocate for access to accurate technology for all. This means speaking out against companies and individuals who exploit technology for their own gain, and pushing for policies and regulations that ensure the benefits of technology are shared equitably. We must also work to educate ourselves and others on the importance of socioectech justice, and take active steps to implement these principles in our own lives.

By coming together and working towards a common goal, we can make the world a better place and ensure that technology is used for the benefit of all, rather than the few. So let us take inspiration from Ella and all the other heroes out there, and work to create a brighter future for everyone through the principles of socioectech justice.

The principles of socioectech justice must be actively pursued and prioritized by individuals, organizations, and communities to ensure that technology is used in an ethical and equitable manner. This is the only way to ensure that the benefits of technology are distributed fairly and that everyone has access to the opportunities that technology provides. As we come to the end of our journey, it is important to remember that the implementation of socioectech justice requires ongoing effort and dedication from everyone involved.

In conclusion, implementing the principles of socioectech justice requires a collective effort from all of us. By working together and taking concrete actions, we can ensure that technology is used in a way that benefits all communities and individuals, regardless of their background or circumstances. And that, my dear students, is a goal worth striving for!

16

Advocating for Access to Technology for All

The advancement of technology has brought about numerous benefits and opportunities for individuals and communities. However, not everyone has equal access to these benefits. The digital divide is a real issue, and it's important to advocate for equal access to technology for all.

In this chapter, we will examine the current state of technology access, the reasons why some individuals and communities do not have access, and ways in which we can work together to bridge the digital divide.

The digital divide in rural America

In rural America, there is a significant divide between the "haves" and the "have-nots" when it comes to technology access. Many rural communities lack access to high-speed internet and up-to-date technology, which makes it difficult for residents to participate in the modern economy and take advantage of online resources. This digital divide is particularly pronounced in low-income and marginalized communities, where residents may not have the resources to purchase or maintain technology, or may not have access to the internet or computers through their jobs or schools.

The digital divide in urban areas

In urban areas, the digital divide can also be a significant issue, with low-income communities often lacking access to the latest technology. This can be due to a number of factors, including a lack of resources to purchase or maintain technology, or limited access to the internet through schools or public spaces. Additionally, many urban communities are gentrifying, which can lead to displacement and the loss of community resources, including technology access. In these cases, it's important for advocates to work together to bridge the digital divide and ensure that all residents have access to technology and its benefits.

The digital divide for people with disabilities

For people with disabilities, access to technology can be a matter of independence and quality of life. Many individuals with disabilities rely on technology to perform everyday tasks, such as communication, navigation, and employment. However, the cost of technology and limited access to assistive technology can be a significant barrier for many people with disabilities. It's important for advocates to work together to bridge this digital divide and ensure that people with disabilities have access to the technology they need to live fulfilling and independent lives.

Another example of the systemic divide is the fragmentation of family law in practice can be seen in the state of Florida. In Florida, the family law court system operates on a county level, with each county having its own budget for funding family law court operations. This has resulted in a wide disparity in the quality and availability of family law resources from county to county.

For example, in one county, the circuit clerk's office may have a comprehensive online application system for filing divorce or child custody cases, while in another county the same process may be done through paper filings and in-person visits to the circuit clerk's office. This disparity can make it difficult for individuals in some counties to access the resources they need to navigate the family law court system, particularly if they have limited access to technology or transportation.

Furthermore, the lack of uniformity in the family law system can also lead to inconsistent application of the law, making it harder for individuals to understand their rights and obligations in family law cases. This fragmentation of the family law system perpetuates the ongoing issues in the field, rather than solving them, causing harm to families and children. It is important for advocates to work together to address these disparities and ensure equal access to technology and resources in all family law court systems.

The fragmentation of family law schools from county board funding budgets and circuit clerk applications of family law in practice is not directly related to the digital divide, but it does compound the various digital divides between different groups of people. Law school students may have access to the latest technology and resources in their education, but when they enter the practical world of family law, they may find that the resources and technology available to them are limited by budget constraints and circuit clerk applications.

Parents in court who need to navigate the family law system may not have access to the same technology or resources that law school students have, leading to a digital divide between those who have access to technology and resources and those who do not. This can make it difficult for parents to understand the legal process and represent themselves effectively in court.

Budget constraints also play a role in the digital divide within the family law system. While law schools may have access to the latest technology, courtrooms and circuit clerks may not have the same resources available to them, leading to a lack of access to technology and resources for those who need it most.

Overall, the fragmentation of family law schools and the limitations on funding for circuit clerks and courtrooms compound the digital divide and make it difficult for those in need of technology and resources to access them. It is important for policymakers and advocates to address these issues in order to ensure that all individuals have equal access to technology and resources in the family law system.

One of the biggest barriers to technology access is cost. Not everyone can afford to purchase a computer, smartphone, or internet service. This puts those who can't afford it at a disadvantage, as they're unable to take advantage of the benefits of technology.

Another barrier to technology access is the lack of digital literacy. Many individuals and communities, especially those in rural or low-income areas, may not have the knowledge or skills to effectively use technology. This can be due to a lack of access to digital literacy programs or a lack of digital education in schools.

We can work to bridge the digital divide by advocating for policies that ensure equal access to technology for all. This can include supporting initiatives that provide affordable or free access to computers and internet service, and programs that teach digital literacy skills.

Additionally, we can also advocate for increased funding for schools in underserved communities, to ensure that all students have access to technology and digital education. This will help close the digital divide and provide equal opportunities for all.

It's important to remember that technology access is a human right. We all deserve equal access to the benefits and opportunities that technology provides. By advocating for access to technology for all, we can help to build a more equitable and inclusive society.

As we move forward in our journey towards a more equitable and just world, it's time to turn our attention towards the next step in the implementation of socioectech justice. In chapter 15, we discussed the importance of advocating for access to technology for all and bridging the digital divide. But now, it's time to focus on driving diversity within the technology industry itself. The technology industry has the power to shape our world and our future, and it's imperative that we ensure that all voices are heard and all perspectives are represented. By promoting diversity in the industry, we can build a more inclusive and equitable future that benefits everyone. With the principles of socioectech justice as our guide, let's work together to drive diversity and bring about a world of freedom, unity, and prosperity for all.

17

Driving Diversity in the Technology Industry

"Diversity is not a problem to be solved, it is a strength to be embraced." - Unknown

As we have seen, the implementation of socioectech justice requires a collective effort from all of us. In order to ensure that technology is used in an ethical and equitable manner, we must work together to promote access to technology for all, advocate for privacy-protective policies, and make conscious decisions about the technologies we use and support. However, this is just the first step in creating a more inclusive and just future. Another critical aspect of this journey is driving diversity in the technology industry itself.

The technology industry has historically been dominated by a narrow segment of the population, primarily white, male, and of a certain socioeconomic background. This has resulted in a homogenized perspective on what technology should look like and what problems it should solve.

This lack of diversity has also led to a lack of representation and a lack of understanding of the diverse needs of communities and individuals. In order to truly create technology that serves the needs of all, we must work to increase diversity in the technology industry and promote a more inclusive culture.

"Inclusion and diversity drive innovation and creativity. The more diverse perspectives we bring to the table, the better the solutions we create."

The inclusive startup

A small startup called InclusivTech was founded with the goal of creating innovative and inclusive technology products. The co-founders, all from diverse backgrounds, made a conscious effort to build a team that was diverse in terms of gender, race, ethnicity, and education. They understood that having a diverse team would lead to better decision-making and create products that would appeal to a wider range of consumers.

InclusivTech was a resounding success. The company created products that were inclusive, accessible, and easy to use. They were able to attract a large and loyal customer base, and the company quickly grew in popularity. Investors were eager to invest in the company, and it was soon valued at over $100 million.

The exclusive corporation

A large corporation called ExclusivTech was known for its cutting-edge technology and high-quality products. The CEO, who was white and male, had a firm belief that the best ideas and solutions came from like-minded people. He was determined to build a team that was exclusively white and male, believing that this was the key to success.

Despite their initial success, the company soon began to suffer. They were unable to attract new customers, and the company's profits began to decline. The CEO was perplexed and could not understand why the company was failing. He refused to acknowledge that the lack of diversity and inclusivity was the root cause of the problem.

As it turned out, the CEO's beliefs were misguided. Consumers were becoming increasingly aware of the importance of diversity and inclusivity, and they were not willing to support companies that did not align with these values. ExclusivTech eventually went bankrupt, and the CEO was left to ponder what might have been if he had listened to the voices of those who understood the importance of diversity and inclusivity.

The two case studies demonstrate the crucial role that diversity and inclusivity play in driving innovation and creativity. InclusivTech's success can be attributed to the fact that its founders understood the importance of a diverse team, and they made a conscious effort to bring together people with different perspectives and experiences. In contrast, ExclusivTech's failure was a result of the CEO's exclusive and narrow-minded approach to building a team. The lesson to be learned is that the more diverse perspectives we bring to the table, the better the solutions we create.

There are many ways to drive diversity in the technology industry, including providing educational opportunities and resources for underrepresented communities, increasing the representation of women and people of color in leadership positions, and fostering a workplace culture that values diversity and inclusivity. By doing so, we can ensure that the perspectives and needs of all communities are represented and addressed in the development of technology.

One successful example of driving diversity in the technology industry is the GirlCode organization, which provides coding and technology education for young girls from underrepresented communities. GirlCode has helped to increase the representation of women in the technology industry and create a more inclusive and equitable future. Additionally, organizations such as Black Girls Code and Code2040 provide similar opportunities for young girls and boys of color to learn coding and technology skills, helping to increase diversity in the industry.

Another example is the conscious effort of tech companies to diversify their hiring practices and workplace culture. Companies such as Google and Apple have implemented initiatives to increase the representation of women and people of color in their workforce and have also established programs to support diversity and inclusivity in the workplace.

However, driving diversity in the technology industry is not just the responsibility of tech companies. It is up to all of us to support and promote diversity and inclusivity in the technology industry. This can be done by supporting organizations that provide educational and career opportunities for underrepresented communities, advocating for policies that promote diversity and inclusivity in the workplace, and by making conscious decisions about which technology companies we support and purchase products from.

Olivia was a single mother of two children who lived in a low-income neighborhood. Despite having a strong passion for technology and a talent for coding, she never had the opportunity to pursue a career in the field. The only technology companies in her area were large corporations that had little to no diversity and inclusivity policies in place. As a result, Olivia and her children were left behind in the digital revolution, unable to access the educational and career opportunities available to those in more affluent areas.

Mrs. Smith was a grandmother who lived in a rural community with her grandchildren. Despite her love for technology, she did not have access to the internet or digital devices, leaving her cut off from the rest of the world. Mrs. Smith's community was also neglected by the technology industry, as large corporations saw no profit in investing in their infrastructure or education.

The lack of diversity and inclusivity in the technology industry perpetuated a cycle of inequality, leaving Olivia, her children, Mrs. Smith and her grandchildren behind in the digital revolution. The technology industry had become a bastion of privilege, perpetuating the divide between the haves and have-nots.

In stark contrast to the lack of diversity and inclusivity in the first case study, a startup technology company led by a group of inclusive and diverse decision makers was changing the game. They had a clear mission to provide educational and career opportunities for underrepresented communities, with a focus on hiring a diverse team and fostering a culture of inclusivity.

The company's products and services were developed with the needs of underrepresented communities in mind, and they provided free digital devices and internet access to low-income families, schools, and community centers. Their commitment to diversity and inclusivity was not just limited to the workplace but extended to the communities they served.

As a result, Olivia and her children were given the opportunity to learn about technology and pursue careers in the field. Mrs. Smith and her grandchildren were given access to the internet and digital devices, enabling them to connect with the world and improve their quality of life. The technology industry was becoming a force for good, breaking down barriers and promoting unity and prosperity for all.

The success of this startup was a testament to the importance of inclusion and diversity in driving innovation and creativity. The more diverse perspectives brought to the table, the better the solutions created, and the greater the impact on society as a whole.

In conclusion, driving diversity in the technology industry is a crucial step in the journey towards socioectech justice. By working together to increase diversity and promote inclusivity, we can create technology that truly serves the needs of all communities and individuals and build a brighter and more equitable future.

As we work to promote diversity and inclusivity in the technology industry, we must also recognize the importance of ethical regulation in the development and use of technology. Just as we strive to ensure that everyone has access to technology, we must also ensure that it is used in a responsible and ethical manner. The next chapter will explore the role of ethical regulation in the technology industry and how we can advocate for it in order to create a better future for all. Together, we can work towards a world where technology is used for the greater good and where everyone has the opportunity to benefit from its advancements. Let us continue to work towards a future of freedom, unity, and prosperity for all.

18

Advocating for Ethical Regulation of Technology

In today's society, technology plays an ever-growing role in our daily lives. From social media to online shopping, technology has made many aspects of life easier and more convenient. However, the rapid advancement of technology has also raised concerns about its impact on our privacy, security, and well-being. It is crucial that we advocate for ethical regulation of technology to ensure that its benefits are shared equitably and its negative consequences are minimized.

The regulation of technology has traditionally been focused on protecting consumers from harmful products and services. However, this approach has proven to be insufficient in addressing the complex challenges posed by technology. The tech industry is constantly evolving, and regulations must keep pace with the latest advancements. This requires a shift from regulating consumers to regulating the producers of technology.

SocioEcTech producers have a responsibility to ensure that their products and services are ethically designed, developed, and marketed. This includes protecting consumer data, ensuring fair pricing, and providing accurate information about the use of their products. SocioEcTech producers must also take into account the broader social and economic impact of their products and services, including the potential for creating digital divides or perpetuating existing inequalities.

Ultra-safe data policies

One example of advocating for ethical regulation of technology can be seen in the efforts to create ultra-safe data policies. With the increasing amount of sensitive information being shared online, it is imperative that technology companies prioritize the privacy and security of their users. In response, organizations have begun implementing measures such as two-factor authentication, encrypted data storage, and strict access controls to ensure that user data remains secure.

Smart and ultra-safe contracts, policies

Another way in which the technology industry is advocating for ethical regulation is through the use of smart contracts. Smart contracts are self-executing agreements that automatically enforce the terms of the contract without the need for intermediaries. This technology is particularly useful in industries such as finance and real estate where contracts are typically complicated and time-consuming to execute. By automating the process, smart contracts help to reduce the risk of human error, increase transparency, and reduce the costs associated with traditional contract execution.

Constructive child safety files in adversarially stacked stages of high conflict corporately systemic divorce

In the context of adversarially stacked stages of high conflict corporately systemic divorce, it is essential to have constructive child safety files. These files provide critical information about the children involved in the divorce, including medical and educational records, as well as information about any custody arrangements or court orders. By implementing policies that ensure the privacy and security of these files, the technology industry can help to prevent sensitive information from being used in a manner that could harm the children involved.

The examples above demonstrate how the technology industry is working to advocate for ethical regulation in order to protect the rights and privacy of its users. Whether it be through ultra-safe data policies, smart contracts, or constructive child safety files in adversarially stacked stages of high conflict corporately systemic divorce, the technology industry is working to ensure that technology is used in an ethical and responsible manner for the benefit of all.

There is no one-size-fits-all solution to regulating technology. The nature of technology and its impact on society are complex and multifaceted. However, a people-centered approach that prioritizes the well-being of communities and individuals is crucial in advocating for ethical regulation of technology. This means considering the social and economic implications of technology, the impact it has on the environment, and the rights of users to privacy and security.

One example of a company and government that have taken a people-centered approach to advocating for ethical regulation of technology is Patagonia and the European Union. Patagonia, the outdoor clothing and gear company, has made a commitment to prioritize the well-being of the environment and the communities in which they operate. This includes using sustainable materials in their products, supporting environmental conservation efforts, and advocating for government policies that protect the environment. The European Union, on the other hand, has implemented the General Data Protection Regulation (GDPR) which prioritizes the rights of users to privacy and security. The GDPR requires companies to obtain consent from users before collecting their personal data and gives them the right to access, erase, and control their personal information. Both Patagonia and the European Union serve as examples of how a people-centered approach to advocating for ethical regulation of technology can lead to positive social and economic outcomes, while also protecting the rights of individuals and the environment.

The tale of two lives

In the small town of Millville, there were two bakeries - one was a small, family-owned bakery that prided itself on using locally sourced ingredients and traditional baking methods to produce high-quality baked goods. The other was a large, multinational bakery chain that relied on industrial-scale production methods and cheaper, mass-produced ingredients.

The family-owned bakery was concerned about the impact their business was having on the environment and local community, so they made a commitment to reducing their carbon footprint and sourcing ingredients from local farmers. They also invested in renewable energy and implemented sustainable business practices, such as recycling and composting.

In contrast, the multinational bakery chain was more concerned with maximizing profits and cutting costs, so they prioritized efficiency and cost-saving measures over ethical considerations. They used cheap, mass-produced ingredients that were sourced from large-scale industrial farms, and they were criticized for their lack of transparency and disregard for the environment and local communities.

As a result, the family-owned bakery was able to establish a loyal customer base and a positive reputation in the community, while the multinational bakery chain struggled to win the trust and support of local customers.

The tale of two communities

In the two cities of Techton and Regulateville, the approach to regulating technology was vastly different. In Techton, the government took a hands-off approach and allowed tech companies to operate with minimal interference, leading to a thriving tech industry but also a growing digital divide and issues with privacy and security.

In Regulateville, the government was more proactive in advocating for ethical regulation of technology. They imposed strict regulations on tech companies to ensure they were protecting user privacy and security, and they also made investments in bridging the digital divide and promoting access to technology for all citizens.

Techton was praised for its tech-friendly environment and dynamic tech industry, but it was also criticized for its lack of regulation and disregard for the well-being of its citizens. Regulateville was praised for its commitment to protecting citizens' rights and promoting social and economic equality, but it was also criticized for being too restrictive and stifling innovation in the tech industry.

In the end, it was clear that balancing the benefits of a thriving tech industry with the need for ethical regulation and consideration for the well-being of citizens was crucial in promoting a better future for everyone.

In other words, it is up to all of us to advocate for ethical regulation of technology. This can be done by supporting organizations that promote responsible technology development, lobbying for regulations that promote ethical practices in the tech industry, and by being conscious consumers who demand products that are ethically designed and marketed. The principles of socioectech justice must guide this movement in order to create a better future for everyone. Together, we can ensure that technology is used for the greater good and that its benefits are shared equitably across all communities and individuals.

With a commitment to ethical regulation of technology, we have taken a vital step in creating a future where all individuals and communities have equal access to the benefits of technology. We have proven that change is possible and that people-centered approaches can drive positive outcomes. Now, it is time to look to the future and be inspired by the success stories of socioectech justice in action. In Chapter 18, "Inspiring Change: Success Stories of socioectech justice", we will explore real-life examples of individuals, communities and organizations who have made a difference and pushed for positive change in the technology industry. These stories of hope and determination will give us the strength to continue our efforts towards a future where technology serves humanity, not the other way around.

19

Inspiring Change: Success Stories of SocioEcTech Justice

The pursuit of socioecotech justice can seem like an overwhelming task. However, there are many inspiring examples of individuals, communities, and organizations that are successfully advocating for a more equitable and sustainable future. These stories provide a source of hope and motivation, and serve as a reminder that we are capable of creating positive change in the world.

One such story is of a small town in rural America that was faced with a digital divide. Despite the rapid growth of technology, many residents in this community did not have access to the internet, which made it difficult for them to participate in the digital economy and access vital resources like healthcare and education. However, this community was not content to be left behind. A group of residents, local businesses, and community organizations came together to create a network of community hotspots, which provided free Wi-Fi to those who could not afford it. The impact of this project was significant, and it helped bridge the digital divide and bring people together in the pursuit of a more equitable future.

Another inspiring example is the work of a tech startup based in Silicon Valley. This company was founded on the principle of ethical technology and social responsibility. The company's products were designed with the user's privacy and security in mind, and they were committed to being transparent and open about the data they collected and how they used it. In an industry where privacy concerns are becoming increasingly prevalent, this company set itself apart by putting the well-being of its users first.

Another example is a school district in New York City that was facing a challenge. Many of its students came from low-income families and did not have access to technology at home. However, the district recognized that technology was becoming an increasingly important tool for learning and success. To address this issue, the district implemented a program that provided students with laptops and internet access. The impact of this program was transformational, and it provided students with a level of access to technology that they never would have had otherwise.

These are just a few examples of the many inspiring stories of socioecotech justice. The individuals, communities, and organizations behind these stories are examples of what is possible when people come together to create positive change in the world. They serve as a reminder that we are all capable of making a difference, and that we have the power to create a more equitable and sustainable future.

In conclusion, the pursuit of socioecotech justice is a noble and important endeavor. However, it can sometimes feel like an overwhelming task. However, these success stories serve as a reminder that we are capable of creating positive change in the world. By learning from these examples and drawing inspiration from them, we can all play a role in inspiring change and advocating for a more equitable and sustainable future.

20

The Emergence of the Digital Empowerment Movement

The digital world has created new opportunities for individuals and communities to connect, learn, and grow. However, this digital landscape has also brought with it a number of challenges, including unequal access to technology, privacy and security concerns, and a lack of ethical regulation. As a result, many individuals and communities have come together to advocate for change and to promote the use of technology in ways that benefit all people.

One example of individuals and communities coming together to advocate for change in technology is the Digital Empowerment Movement. The Digital Empowerment Movement is a global movement that is working to bridge the digital divide and promote digital literacy, digital inclusion, and digital rights for all. This movement was started by a group of activists, tech experts, and community leaders who saw the potential for technology to empower communities and make a positive impact in people's lives.

The Digital Empowerment Movement has launched various initiatives and programs aimed at increasing access to technology and digital skills, especially among marginalized communities. For instance, they have set up community technology centers in low-income neighborhoods, where people can access computers and the internet for free or at a low cost. They also offer digital literacy classes, workshops, and mentorship programs to help people acquire the skills they need to fully participate in the digital economy.

The Digital Empowerment Movement is an emerging social and political movement that aims to ensure that technology is used in ways that are inclusive, equitable, and responsible. This movement is made up of individuals, organizations, and communities that are committed to promoting the use of technology to enhance the well-being of people and communities.

At its core, the Digital Empowerment Movement seeks to ensure that technology is accessible, secure, and empowering for everyone, regardless of their background, income, or location. The movement is founded on the belief that technology should be used to enhance the quality of life for all people, rather than being used as a tool for oppression or exploitation.

To achieve these goals, the Digital Empowerment Movement is focused on three main areas: advocacy, education, and community building. Through advocacy, the movement seeks to promote policies and initiatives that ensure that technology is used in ways that are equitable, inclusive, and responsible. This includes advocating for access to technology for all individuals, regardless of their income or location, and advocating for policies that protect the privacy and security of individuals in the digital world.

The Digital Empowerment Movement has been at the forefront of advocating for change in the technology industry. They are raising awareness about issues such as the digital divide, unequal access to technology, and the need for ethical regulation of technology. One example of this is the "Save the Internet" campaign, where activists came together to protest against the repeal of net neutrality laws. The campaign was successful in bringing attention to the issue and pressuring lawmakers to reconsider their stance on net neutrality.

Education is another key area of focus for the Digital Empowerment Movement. This involves educating individuals and communities about the benefits and risks of technology, and empowering them with the knowledge and skills they need to make informed decisions about their use of technology. This includes initiatives such as digital literacy programs, training programs for community leaders, and workshops on digital privacy and security.

The Digital Empowerment Movement has been at the forefront of advocating for change in the technology industry. They are raising awareness about issues such as the digital divide, unequal access to technology, and the need for ethical regulation of technology. One example of this is the "Save the Internet" campaign, where activists came together to protest against the repeal of net neutrality laws. The campaign was successful in bringing attention to the issue and pressuring lawmakers to reconsider their stance on net neutrality.

Finally, community building is a critical component of the Digital Empowerment Movement. This involves bringing individuals and communities together to share their experiences and to work together to create positive change. This can include initiatives such as local community events, online forums, and social media groups that bring together individuals who are passionate about using technology for good.

The Digital Empowerment Movement has been at the forefront of advocating for change in the technology industry. They are raising awareness about issues such as the digital divide, unequal access to technology, and the need for ethical regulation of technology. One example of this is the "Save the Internet" campaign, where activists came together to protest against the repeal of net neutrality laws. The campaign was successful in bringing attention to the issue and pressuring lawmakers to reconsider their stance on net neutrality.

Everyone has the potential to be empowered, even from the comfort of their own home. This can be done through various means such as:

1. Advocacy: By advocating for changes in technology and its regulation, individuals can play a role in shaping the direction of technology and how it is used. This can be done by participating in advocacy groups, writing to elected officials, and supporting organizations that promote responsible technology use.

2. Education: By learning about technology, its impact on society, and how it can be used for good, individuals can become more informed and empowered to use technology in a way that benefits everyone. This can be done by taking online courses, attending workshops or conferences, and reading books and articles on technology and its impact.

3. Community building: By connecting with others who share a similar passion for technology and its impact, individuals can build a strong community of like-minded individuals. This can be done by attending community events, joining online forums or groups, and participating in local initiatives that promote responsible technology use.

By focusing on these three main areas, individuals can play a role in promoting the use of technology in ways that benefit all people, and empower themselves and their communities to drive positive change.

The Digital Empowerment Movement is a powerful force for positive change, and it is inspiring individuals and communities all over the world to work together to create a more equitable and inclusive digital landscape. Whether through advocacy, education, or community building, the Digital Empowerment Movement is helping to ensure that technology is used in ways that benefit all people, now and in the future.

21

The Rise of Social Entrepreneurship in Tech

"Social entrepreneurship is not just about creating profitable businesses, but about using technology and innovation to solve societal problems and make the world a better place."

The rise of social entrepreneurship in tech has been a significant development in recent years. Social entrepreneurship is the practice of combining the traditional principles of entrepreneurship with the goal of addressing social and environmental issues. With technology playing an increasingly dominant role in our daily lives, it makes sense that social entrepreneurs are leveraging technology to tackle some of the world's biggest problems.

One of the key drivers of social entrepreneurship in tech is the need for solutions that address the digital divide and make technology more accessible to all. In addition, many social entrepreneurs are working to use technology to address issues such as income inequality, environmental degradation, and access to healthcare.

"Technology and social entrepreneurship go hand in hand. By harnessing the power of technology, social entrepreneurs can reach more people, have a greater impact, and create sustainable solutions to social and environmental problems."

One of the most important aspects of social entrepreneurship in tech is the focus on finding sustainable solutions that benefit both society and the environment. Social entrepreneurs are using technology to create products and services that are not only innovative, but also have a positive impact on the world. This requires a deep understanding of the issues they are addressing, as well as a willingness to collaborate with others to create solutions that are both effective and scalable.

The organization "Kiva" is a prime example of social entrepreneurship in tech. Kiva leverages technology to connect people from all over the world in support of small business owners and entrepreneurs in developing countries, providing them with micro-loans to help them grow their businesses and improve their livelihoods.

A social entrepreneur named Jane started a tech company that provides affordable and accessible healthcare services to underserved communities in rural areas. She saw a gap in healthcare access and decided to use technology to bridge that gap. She developed a telemedicine platform that allows patients to connect with doctors and other healthcare professionals from the comfort of their own homes. With her technology, patients can receive medical advice and treatment without having to travel long distances to see a doctor. In addition to improving access to healthcare, Jane's company has created job opportunities and helped to spur economic growth in rural communities.

"Girls Who Code" is another example of social entrepreneurship in tech. This organization aims to close the gender gap in technology by providing girls and young women with the education and resources they need to pursue careers in computer science and technology.

Another important aspect of social entrepreneurship in tech is the focus on education and training. Many social entrepreneurs are working to provide educational opportunities and training programs to help people acquire the skills they need to succeed in a rapidly changing technological landscape. This includes programs that teach coding and other technical skills, as well as programs that help people learn how to use technology in a more responsible and sustainable way.

The rise of social entrepreneurship in tech has the potential to revitalize socioectech justice for all in several ways. First, social entrepreneurs are driven by a desire to make a positive impact in the world, rather than solely by profit. This means that they are more likely to prioritize the needs of communities and individuals, rather than just the bottom line.

Secondly, social entrepreneurs are often more innovative in their approach to technology and problem-solving. They are not limited by traditional thinking or corporate bureaucracy, and are able to take risks and experiment with new ideas in order to find solutions to pressing problems. This leads to the development of new technologies and applications that can improve the lives of individuals and communities.

Finally, social entrepreneurs often work in close collaboration with communities and stakeholders, rather than simply imposing solutions from the top down. This ensures that their efforts are aligned with the needs and values of those they serve, and that the benefits of technology are more widely shared.

Overall, the rise of social entrepreneurship in tech is a powerful force for change, and can help to revitalize socioectech justice for all by creating innovative, people-centered solutions to the challenges facing our world today.

As we've seen, the rise of social entrepreneurship in tech has been a game changer in the world of socioecotech justice. The power of individuals and communities coming together to make a difference is truly inspiring. But what if I told you that there's an even greater way to drive change and create impact? Let's explore the power of community technology hubs in chapter 21. Imagine a world where everyone, no matter their background or socioeconomic status, has access to the resources and support they need to succeed in the digital age. That world is within our reach, and community technology hubs are making it a reality. So get ready to be inspired as we delve into this exciting new chapter!

22

The Power of Community Technology Hubs

Technology has changed the way we live, work, and communicate. In today's fast-paced world, access to technology has become a fundamental right. Unfortunately, many communities are still left behind, struggling to access and make use of technology. This is where community technology hubs come into play. These hubs are community-based spaces where people can access technology, learn new skills, and connect with others.

In a small rural town in the southern United States, a young girl named Maria was growing up without access to technology. Despite her family's best efforts, they could not afford a computer or internet access, making it difficult for Maria to keep up with her studies. She was often teased by her classmates who had the latest gadgets and technology, leaving her feeling left out and inadequate.

Maria's school had limited resources, and although they had a computer lab, it was only open a few hours a week, and the outdated machines were slow and often didn't work. As a result, Maria was struggling to keep up with her school work and was falling behind. She was losing her passion for learning and felt hopeless about her future.

Despite her situation, Maria was determined to succeed. She would walk to the local library every week to use their computers, but it was not enough. She needed more support and resources to help her reach her full potential. Her community was filled with families who were in similar situations, and there was a clear need for a community technology hub.

Unfortunately, there was not enough funding or resources available to establish such a hub in the area. Maria's dream of accessing technology and improving her education was slipping away. This is a sad but all too common story for many communities who are still left behind, struggling to access and make use of technology.

Community technology hubs are changing the lives of people in many communities. They provide access to technology and the Internet, training, and resources to help people learn new skills, and opportunities to connect with others. The hubs are staffed by trained volunteers, who provide one-on-one assistance and support to users. The hubs also provide access to technology tools, such as computers, printers, and other digital devices, as well as software and online resources.

One of the key benefits of community technology hubs is that they bring together people from diverse backgrounds and ages to work together and support each other. This creates a sense of community and promotes social cohesion. Hubs also provide a safe and welcoming environment for people to explore and learn about technology. They are an excellent resource for young people, who can benefit from learning about technology, developing new skills, and connecting with others.

Chris, a resident of a low-income community, had always struggled with access to technology and the Internet. He had never been able to afford a computer, and the local library only had a limited number of computers available for use. This made it difficult for Chris to search for job opportunities, access educational resources, and connect with friends and family.

But everything changed when a community technology hub opened in his neighborhood. Chris was able to access a computer, free of charge, and was introduced to a world of opportunities. He received training on basic computer skills, and was able to brush up on his resume and job search skills. With the help of the trained volunteers, Chris was able to secure a job within a few weeks.

Chris was also able to connect with others in the community and build new relationships. The hub provided a safe and welcoming environment where people of all ages and backgrounds could come together and learn from each other. Chris was able to share his experiences with others and help them with their own job search and technology skills.

The community technology hub changed Chris's life for the better. He was able to improve his job prospects, connect with others, and learn new skills. The hub provided a beacon of hope and a place of community in his neighborhood. It showed him the power of technology to bring people together and empower individuals to reach their full potential.

Community technology hubs also play a crucial role in providing access to technology and digital resources to people who may not have access otherwise. They serve as a bridge between the digital divide and provide access to technology to people who would not have it otherwise. These hubs help to close the gap between those who have access to technology and those who do not.

In Central, Illinois, there was a growing homeless population who struggled to access basic resources like food, shelter, Bible study and healthcare. However, one community technology hub made a difference in their lives by providing access to technology and the Internet.

At the hub, the homeless were able to connect with organizations and resources that could help them get back on their feet. They were able to use the computers to search for job opportunities and housing options, and access online resources for healthcare and other basic needs. The hub also offered training and workshops on how to use technology and the Internet, which helped the homeless gain valuable skills and knowledge.

One man named Tom was homeless for several years and had given up hope of ever finding a job or a place to call home. However, after visiting the community technology hub, he was able to use the computers to search for job opportunities and found a job within a week. He was then able to save enough money to rent a small apartment and start a new life.

The community technology hub not only provided access to technology and the Internet, but it also brought people together and created a sense of community. The homeless felt welcomed and supported, and the staff and volunteers provided them with one-on-one assistance and support.

The success of the community technology hub in helping the homeless has inspired other cities to set up similar initiatives. Access to technology and the Internet has changed the lives of many people in this community, providing them with the tools and resources they need to get back on their feet and start a new life.

Community technology hubs have the power to change lives and bring people together. They provide access to technology and digital resources, opportunities to learn new skills, and a safe and welcoming environment for people to explore and connect with others. As we continue to live in a rapidly changing digital world, community technology hubs will play an increasingly important role in promoting socioectech justice for all.

As we have seen in the previous chapter, community technology hubs have become a powerful tool for promoting socioectech justice and providing access to technology for all. However, despite the positive impact that these hubs have had, there are still many barriers that need to be overcome in order to fully realize the potential of socioectech justice. In this next chapter, we will explore these barriers and discuss strategies for overcoming them, so that we can continue to make progress towards a more inclusive, equitable and prosperous society.

23

Navigating, Overcoming and Adapting To Barriers to Implement SocioEcTech Justice

"Privacy is the power to selectively reveal oneself to the world." - Ester Dyson

As we move forward in creating a more inclusive and just society, it is important to acknowledge and understand the barriers that exist to implementing socioecotech justice. From economic disparities to lack of access to technology, these barriers can often be significant and challenging to overcome. However, with determination, creativity, and collaboration, it is possible to navigate these barriers and create a more equitable future for all.

One of the main barriers to socioecotech justice is the digital divide, which refers to the gap between those who have access to technology and those who do not. This divide often intersects with other forms of inequality, such as socioeconomic status, race, and geographic location. To overcome this barrier, it is important to increase access to technology and digital literacy programs in underrepresented communities. This can be done through initiatives such as community technology hubs, digital inclusion programs, and digital skills training.

The ego gap is the gap between powerful people who are overly focused on their own interests, and those who are not. This can result in a lack of consideration for others, and a failure to understand the impact that their actions have on the wider community. This can be seen in various ways, such as the use of technology to gather and use data without consent, or the creation of policies that restrict access to technology for certain groups of people. When powerful individuals are more concerned with their own interests than the well-being of others, they can be seen as denying others the ability to achieve socioectech justice. This can be seen as a major barrier to implementing socioectech justice and must be overcome in order to create a more inclusive and equitable society.

"Security is both a means and an end. It is essential in order to protect the freedoms we enjoy." - Benazir Bhutto

One example of an ego gap getting in the way of navigating, overcoming and adapting to barriers is in the world of politics. There was a time when a certain politician was running for office and they believed that they were the only one who could solve the problems of the country. They were so focused on themselves and their own interests that they completely disregarded the needs and concerns of the people they were supposed to represent.

This politician was so blinded by their ego that they refused to listen to the input of others, even those who had years of experience and expertise in the areas they were trying to address. They were so convinced that they were right that they disregarded the opinions of others and made decisions that had negative consequences for the people they were supposed to serve.

In this case, the ego gap between the politician and the people caused major barriers to implementing socioectech justice. The politician's refusal to listen to the needs of the people and their inability to adapt to new ideas and perspectives created a major barrier to progress and hindered the implementation of policies that would have benefited the people.

One example of an ego gap causing a major barrier in navigating a binding policy adverse to those who had to pay for what another adversary thought instead of real-time data is the case of a large corporation implementing a new employee benefits package. The CEO of the corporation, who had a large ego, believed that they knew what was best for the employees and did not take into consideration the actual needs and wants of the employees. They implemented a benefits package that was costly for the employees, and did not meet the needs of the majority of the employees.

The employees, who were not consulted during the policy making process, were frustrated and felt unheard. They expressed their concerns, but the CEO was too absorbed in their own thoughts and did not take the time to listen to the employees. As a result, the employees felt disrespected and disengaged, which resulted in high turnover rates and low morale in the workplace.

This case is a clear example of how an ego gap can cause a major barrier in navigating a binding policy adverse to those who had to pay for what another adversary thought instead of real-time data. It highlights the importance of considering the perspectives and opinions of all stakeholders when making important decisions, rather than relying solely on the opinions of those in power.

An example of an ego gap being a major barrier in overcoming a binding policy that was adverse to those who had to pay for what another adversary thought, instead of relying on real-time data, can be seen in the case of the introduction of a new tax policy in a certain country. The government, without proper research and analysis, implemented a new tax policy that was heavily biased towards large corporations, who were believed to have more resources and capability to pay the additional tax. However, small businesses, who were the backbone of the country's economy, were greatly affected by this policy. They were forced to pay more tax, which significantly impacted their ability to sustain their business operations.

Despite the small businesses raising their concerns and presenting real-time data that showed the adverse impact of the policy on their operations, the government, driven by their own ego and perception of the situation, refused to listen. They believed that the small businesses were not capable enough to understand the complexity of the policy and thus, ignored their cries for help.

However, as the small businesses continued to suffer and the economy of the country started to deteriorate, the government was eventually forced to revisit the policy and take the real-time data into consideration. The ego gap that existed between the government and the small businesses was a major barrier in overcoming the policy and implementing a more inclusive and effective solution.

One example of this kind of ego gap can be seen in the situation of a corrupt leadership taking bribes during the Covid-19 pandemic. This leadership was heavily influenced by the opinions of previous adversaries and disregarded real-time data and facts, which resulted in a binding policy that was adverse to those who were already struggling to make ends meet. The ego gap between the corrupt leaders and the people was a major barrier in adapting to the policy and finding a solution that would benefit everyone, not just a select few. Despite numerous protests and calls for action, the corrupt leaders refused to change their stance and instead pushed forward with their self-serving policies, further exacerbating the suffering of the community. This unfortunate situation highlights the dangerous consequences of ego gaps and the importance of staying grounded in reality and making decisions based on facts, not just opinions and personal biases.

Ego gaps can be a major barrier to navigating, overcoming and adapting to barriers in implementing socioectech justice. It is important for leaders to be humble and open-minded, to listen to the needs and concerns of the people, and to be willing to work together to find solutions that will benefit everyone.

Another major barrier to socioecotech justice is the issue of privacy and security in the digital age. As more and more of our personal and sensitive information is stored online, it becomes increasingly important to protect this information from theft, misuse, and abuse. To address this, we must advocate for strong privacy laws, educate people about how to protect their personal data, and work to create secure and transparent technology systems.

The issue of privacy and security in the digital age is a major barrier to socioecotech justice. In a world where technology is constantly evolving, individuals and organizations are concerned about the protection of their personal information and sensitive data. This concern is valid and it is important to protect privacy and security in the digital age. However, the issue of privacy and security can be twisted to fit the interests of those in power, rather than addressing the real concerns and experiences of those who are affected.

For example, governments and corporations may use the guise of privacy and security to justify surveillance and control of citizens and consumers. This can lead to a violation of privacy rights and the collection of sensitive information without proper consent. Additionally, it can result in the suppression of freedom of speech and the suppression of dissent. This is a clear example of how the issue of privacy and security can be twisted to benefit those in power and perpetuate socioecotech injustice.

On the other hand, it is important to find a balance between privacy and security and the protection of citizens and consumers. This can be achieved through the implementation of ethical regulations and standards that prioritize the privacy rights and safety of individuals, while also ensuring that technology is used for the greater good. This includes providing transparency about how personal information is collected, used, and stored, and giving individuals control over their own data.

The issue of privacy and security in the digital age is a major barrier to socioecotech justice. However, it can also be an opportunity to promote ethical regulations and standards that prioritize the privacy rights and safety of individuals. By addressing these concerns, we can create a more inclusive and equitable digital society that benefits all.

There are also obstacles to implementing socioecotech justice at the corporate and government levels. The influence of powerful corporations and vested interests can make it difficult to prioritize the well-being of individuals and communities. To overcome this, we must work to increase transparency, promote accountability, and create regulations that prioritize the needs of people over the interests of corporations.

Finally, there are challenges to implementing socioecotech justice related to the speed and pace of technological innovation. As technology continues to evolve rapidly, it can be difficult to keep up and ensure that policies and practices are in place to promote justice and equality. To address this, we must foster open and inclusive dialogue between technology experts, policymakers, and communities to ensure that everyone has a voice in shaping the future of technology.

In conclusion, there are numerous barriers to implementing socioecotech justice, but with a commitment to social and technological progress, these barriers can be overcome. Through education, community building, and advocacy, we can create a future where technology serves the needs of all people and promotes equality, freedom, and prosperity for all.

As we delve into the challenges and obstacles of implementing socioecotech justice, it's important to remember that progress is possible. By working together and facing these barriers head on, we can create a more equitable and sustainable future for all. In the next chapter, we'll explore the steps we can take to ensure that socioecotech justice remains a driving force for good, long into the future.

24

Ensuring Sustainability for All through SocioEcTech Justice

"Sustainability is not just about the environment, it's about social and economic sustainability as well."

The world is constantly evolving and changing, but one thing remains constant - the importance of sustainability. In today's fast-paced and rapidly advancing world, technology is playing an increasingly crucial role in achieving sustainability. This is where socioecotech justice comes into play. It is a means of ensuring that technology is used in ways that benefit all people, regardless of their background or circumstances. This is especially important when it comes to the use of technology in environmental and social sustainability.

"Technology is the answer to many of our sustainability problems, but only if it is used wisely."

Smart energy systems are a prime example of how technology can be used in a wise and sustainable manner. These systems gather real-time data from various sources and use this information to optimize energy usage and minimize waste. For example, by monitoring energy usage patterns in buildings, the system can adjust heating and cooling systems to use less energy during peak times. Additionally, smart energy systems can also monitor and control energy usage in public transportation, helping to reduce emissions and lower the carbon footprint of the city.

By reducing waste and increasing efficiency, smart energy systems help cities to become more sustainable and environmentally friendly. Furthermore, they can also help to lower energy costs for residents, making it easier for them to access and use energy resources. This, in turn, promotes socioecotech justice, as all members of the community can benefit from the benefits of technology and sustainability.

Smart energy systems are just one example of how technology can be used to promote sustainability and socioecotech justice. There are many other technologies and innovations that can be implemented in similar ways, providing benefits for all members of the community. Through continued investment in these technologies and a commitment to sustainability, we can ensure that everyone has access to the resources and opportunities they need to thrive and reach their full potential.

For the past decade, Dr. Samantha Green, a renowned college professor, has been educating students on the topic of sustainability. With a passion for environmental preservation and a deep understanding of the crucial role technology plays in the sustainability movement, Dr. Green has been incorporating technology into her teaching methods. Her students have the opportunity to learn about the latest sustainable technologies and are taught how to evaluate their efficiency and potential for widespread use.

Throughout the course, Dr. Green emphasizes the importance of hands-on experience and group discussions. This approach allows her students to gain a deeper understanding of the impact technology can have on creating a more sustainable future. Through her teachings, Dr. Green hopes to inspire the next generation of sustainability leaders and empower them to make a positive impact on the world through the use of technology.

Not only does Dr. Green's course provide students with the knowledge and skills necessary to create a more sustainable future, but it also instills a sense of hope and inspiration. Her students leave the course with a greater understanding of the potential for technology to bring about positive change, and with a commitment to utilizing their skills and knowledge to contribute to a better world.

Dr. Samantha Green's passion for sustainability and technology has not only been inspiring to her students, but to everyone who has the privilege of learning from her. She shows us that with a combination of education and technology, we can create a better future for generations to come.

An example of using technology wisely for sustainability is the implementation of smart energy systems in Amsterdam, Netherlands. The city has implemented a comprehensive smart energy system that uses real-time data to optimize energy usage and reduce waste. The system connects all the buildings in the city and allows them to share energy, leading to more efficient energy usage and reducing the city's carbon footprint. The city's smart energy system has already reduced energy usage by 30% and has become a model for other cities around the world to follow. The system's success can be attributed to the collaboration between the city government, private sector, and residents, who have all come together to promote sustainable energy practices. This case study shows that when technology is used wisely and in collaboration with different stakeholders, it can lead to significant benefits for the environment and the community.

Another example of using technology for sustainability is the widespread adoption of electric and hybrid vehicles. These vehicles reduce our dependence on fossil fuels and decrease carbon emissions, leading to a more sustainable transportation system.

Regenerative filing and data recovery networks refer to a system that ensures the protection and preservation of important data and files, especially in situations where they may be at risk of being lost or corrupted. This is particularly relevant in the context of child safety files, where records of children's ultra-safety and well-being are critical for their protection and growth.

Unfortunately, in many cases, ego-centric adult fighting file cases leading up to contracts can cause file storms that put these important files at risk. This can occur when individuals prioritize their own interests over the well-being of children and the integrity of their safety files. For example, some may intentionally manipulate or delete information to advance their own agendas, while others may fail to properly secure or backup these files, leading to accidental loss or corruption.

To overcome these challenges, regenerative filing and data recovery networks can play a crucial role in ensuring the sustainability and protection of child safety files. These networks use advanced technology to automate backup and recovery processes, ensuring that important data is securely stored and easily recoverable in the event of any data loss or corruption. Additionally, these networks also incorporate robust security measures to prevent unauthorized access or tampering, ensuring that the files are protected from potential ego-centric adult fighting.

Overall, regenerative filing and data recovery underground network represents an important step towards ensuring sustainability and protecting the rights and welfare of children in the face of adult ego-centric fighting.

By providing reliable and secure data storage, these networks help to safeguard the well-being and future of our next generation.

Sustainability is about more than just the environment; it is about creating a world that is equitable and just for all people. Technology can help to achieve this by providing access to information, resources, and opportunities that can support and enhance sustainability efforts. Socioecotech justice is crucial in ensuring that the benefits of technology are not just enjoyed by a few but are shared by all.

To ensure sustainability for all, it is important to understand the complex interplay between technology, society, and the environment. This requires a multi-disciplinary approach that considers the social, economic, and environmental impacts of technology. This is where socioecotech justice can make a significant difference.

For example, consider the use of renewable energy sources like solar and wind power. These technologies can provide clean, sustainable energy that can help to reduce our reliance on fossil fuels and mitigate the impacts of climate change. However, to ensure that these technologies are used in a way that is sustainable for all, it is important to consider the social and economic impacts of their use. Socioecotech justice can help to ensure that the benefits of renewable energy are shared by all, regardless of their background or circumstances.

In conclusion, socioecotech justice is a critical component of ensuring sustainability for all in the digital age. It requires a commitment to equity, justice, and fairness in the use of technology. By working together to promote socioecotech justice, we can create a world that is sustainable, equitable, and justice for all.

As we come to the close of chapter 23, it is time to reflect on the progress we have made in ensuring sustainability for all through SocioEcTech Justice. The advancements in technology have given us the tools to build a future that is more sustainable, equitable, and prosperous. Now, it is time for us to come together and build a movement for inclusive socioeconomics and SocioEcTech Justice. Let us use this moment to unify our voices and work towards a future where every person has access to the resources and opportunities they need to thrive. Let us embrace the power of technology to create a world of freedom, unity, and prosperity for all.

25

A Stance for Unity and SocioEcTech Justice

In order for socioectech justice to be fully realized and sustained, it is essential that a movement is built around the principles of inclusive socioeconomics. This framework should be based on the idea of creating a fairer and more equitable society for all, where technology and innovation serve the interests of the people and not just those in positions of power.

To stand in this movement, a number of key steps need to be taken. Firstly, education and awareness must be promoted on a wide scale. This includes educating people about the benefits and potential of socioectech justice, and how it can help to create a more sustainable and inclusive future. This can be achieved through a combination of workshops, events, and other outreach initiatives.

The "SocioEcTech Justice Roadshow"

Meet Tom and Amanda, two recent graduates who were passionate about promoting the message of socioectech justice. After taking a course on sustainability, they realized the importance of spreading awareness about the topic and decided to take action.

Tom and Amanda decided to embark on a nationwide road trip, visiting colleges, universities, and community centers to hold workshops and events about socioectech justice. They dubbed their journey the "SocioEcTech Justice Roadshow".

At each stop, Tom and Amanda set up interactive and entertaining displays, highlighting the benefits of SocioEcTech Justice and how it can help create a more sustainable and inclusive future. They also held workshops where attendees could participate in hands-on activities and group discussions, allowing them to learn more about the topic in a fun and engaging way.

One of their most memorable stops was at a local elementary school, where they hosted a workshop for students. The children were captivated by the interactive displays and hands-on activities, and were eager to learn about the importance of sustainability and how they could help make a difference.

As their roadshow progressed, Tom and Amanda continued to receive positive feedback from those they reached, inspiring them to continue their mission. By the end of their journey, they had visited over 50 cities and reached thousands of people, inspiring a new generation of individuals to stand up for socioectech justice.

The "SocioEcTech Justice Roadshow" was a prime example of how education and awareness can be promoted in a fun and entertaining way. By spreading the message in a creative and engaging way, Tom and Amanda were able to make a real impact, inspiring others to take action and stand up for socioectech justice.

In addition, advocacy and action must be taken to bring about real change that lasts. This includes advocating for policies and initiatives that support socioectech justice, and engaging in acts of resistance against those that undermine it. It also involves the creation of new institutions and systems that support the principles of inclusive socioeconomics and socioectech justice.

One such leader is Jane Doe, a social entrepreneur who has dedicated her career to promoting inclusive socioeconomics and socioectech justice. Jane has been a vocal advocate for policies that support these principles, working with local and national organizations to create programs that empower communities and drive positive change. In addition to her advocacy work, Jane has also established a number of organizations and initiatives that support inclusive socioeconomics and socioectech justice. These include community-based micro-lending programs, technology training centers, and a platform that connects businesses and individuals to eco-friendly suppliers and products.

Through her tireless efforts, Jane has inspired countless others to join the movement for inclusive socioeconomics and socioectech justice. She has shown that it is possible to make a real difference, and that through collaboration and a commitment to positive change, we can build a better future for all. Whether it is through her speaking engagements, community outreach programs, or through the innovative organizations she has established, Jane's work serves as a powerful example of how one person can make a lasting impact in the fight for socioecotech justice.

Another important aspect of building a movement is the establishment of partnerships and collaborations with like-minded organizations and individuals. This includes forming alliances with organizations working on related issues, such as environmental protection, social justice, and human rights, in order to promote a common agenda and leverage collective resources.

A prime example of a successful partnership in the movement for inclusive socioeconomics and socioectech justice is the collaboration between the non-profit organization "Green Future" and the tech company "EcoTech Solutions." Green Future, a non-profit dedicated to promoting sustainability and eco-justice, recognized the critical role that technology plays in creating a more sustainable future. To bring their vision to life, they partnered with EcoTech Solutions, a company specializing in developing sustainable technology solutions.

Together, Green Future and EcoTech Solutions launched a number of projects aimed at promoting the adoption of sustainable technology and raising awareness about the importance of socioectech justice. They organized workshops, events, and outreach initiatives to educate the public about the potential of sustainable technology and the benefits of socioectech justice. They also advocated for policies that support socioectech justice and took action against those that undermine it.

The partnership between Green Future and EcoTech Solutions serves as a model for others seeking to build a movement for inclusive socioeconomics and socioectech justice. By coming together and leveraging their strengths and resources, they were able to have a much greater impact than they could have had alone. Their collaboration has helped to build momentum for the movement and has inspired others to join the cause.

Lastly, a strong and inclusive leadership must be fostered within the movement. This includes developing leaders from diverse communities and backgrounds, who can help to shape the direction of the movement and ensure that the voices of those most affected by socioeconomic and technological inequalities are heard.

By taking these steps and working together, a movement for inclusive socioeconomics and socioectech justice can be built and sustained. With this movement, we can create a future that is more equitable, sustainable, and inclusive for all.

"The future is not something we enter. The future is something we create."
- Leonard I. Sweet

26

Conclusion: The Future of Inclusive Socioeconomics and the Vital Role of SocioEcTech Justice

The world we live in today is facing numerous challenges that threaten to undermine the sustainability and prosperity of our communities and future generations. From economic inequality and social injustice, to environmental degradation and technological advancements that leave many behind, the need for a new paradigm for inclusive socioeconomics and socioectech justice is more pressing than ever.

Over the course of this book, we have explored the concept of inclusive socioeconomics, and how it can be applied to promote equity, sustainability, and resilience in our communities. We have also examined the crucial role that technology can play in this movement, through the development of innovative solutions that support inclusive socioeconomics and help to create a more sustainable and equitable future.

We have delved into the idea of inclusive socioeconomics and how it can drive forward equity, sustainability, and resilience within our communities. By focusing on the intersection of economics and society, inclusive socioeconomics seeks to create a more just and equitable future, where everyone has access to the resources and opportunities they need to thrive.

At the core of this movement is the idea of socioectech justice, which recognizes the vital role that technology can play in promoting sustainability and equity. By harnessing the power of technology, we can create innovative solutions that support inclusive socioeconomics and help to build a more sustainable, resilient, and equitable future for all.

From exploring the benefits of renewable energy systems and smart cities, to examining the impact of electric and hybrid vehicles on sustainability, to discussing the importance of regenerative filing and data recovery networks for child safety files, this book has highlighted the numerous ways in which technology can be used to promote a more sustainable and equitable future.

Additionally, we also looked at the importance of education and awareness, advocacy and action, and partnerships and collaborations in building a movement for inclusive socioeconomics and socioectech justice. By working together and taking a stand, we can create real, lasting change and build a brighter, more equitable future for all.

This book has shown us that inclusive socioeconomics and socioectech justice are critical components of building a sustainable and equitable future. By recognizing the vital role that technology can play in this movement, we can harness its power to create innovative solutions that promote sustainability, equity, and resilience for all.

At the heart of this sustainability is a commitment to ensuring that everyone, regardless of their background, has equal access to the opportunities and benefits of technology. This requires a shift away from the traditional model of technological development, which is driven by profit and focused on the needs of a select few. Instead, it calls for a new approach, one that is centered around the needs and well-being of all people, and that recognizes the interconnections between technology, society, and the environment.

Building a fluid frame for inclusive socioeconomics and socioectech justice requires collaboration and partnership among individuals, organizations, and institutions from all sectors of society. This includes businesses, governments, NGOs, and individuals working together to promote education and awareness, advocate for policies that support this movement, and take action to bring about lasting change.

In order to promote a more inclusive and sustainable future, we need to shift our focus from profit-driven technological development to a model that prioritizes the well-being of all people. This means creating opportunities for everyone to have equal access to the benefits of technology, regardless of their background. To achieve this, we need collaboration and partnership across all sectors of society, including businesses, governments, NGOs, and individuals.

Examples of initiatives that support this movement include education and awareness campaigns, advocacy for policies that support socioectech justice, and the creation of new institutions and systems that prioritize inclusive socioeconomics. Nonprofits and businesses can play a key role by partnering with like-minded organizations and individuals to build custom-tailored movements. These initiatives aim to promote a more sustainable and equitable future, by ensuring that everyone has equal access to technology and its benefits.

Ultimately, building a fluid framework for inclusive socioeconomics and socioectech justice requires a commitment from all of us to work together towards a common goal. By fostering collaboration and partnership across sectors, we can create a future where everyone has the opportunity to thrive, regardless of their background or circumstances.

The future of inclusive socioeconomics and socioectech justice is in your hands, and we must work together to ensure that it becomes a reality. With the right tools, partnerships, and determination, we can create a world that is more equitable, sustainable, and resilient for all. The time for action is now, and the journey towards a better future starts with each one of us.

"The future is not just something we inherit, it's something we create and add value to. And in your hands lies the power to shape it into a more inclusive, sustainable and just world."

27

A Call to Action for a More Just and Inclusive Tech-Enabled Society

In recent years, technology has revolutionized the way we live, work, and interact with each other. It has brought us endless possibilities and endless opportunities. However, as we have explored in this book, the benefits of technology are not equally distributed.

There are those who are left behind and excluded from the technological revolution. There are those who suffer from the negative impacts of technology, such as digital inequalities, job displacement, and environmental degradation.

It is clear that we need to do better. We need to create a more just and inclusive tech-enabled society, one that values and supports everyone, regardless of their background or circumstances. To achieve this, we need to take action, and we need to do it now.

First and foremost, we need to prioritize education and awareness. We need to educate ourselves, our communities, and our leaders about the importance of inclusive socioeconomics and socioectech justice. We need to raise awareness about the challenges that exist, and the solutions that can help to address them.

Second, we need to advocate for policies and initiatives that support this movement. We need to speak out against those that undermine it, and we need to engage in acts of resistance against those that threaten to harm our communities.

Third, we need to collaborate and partner with like-minded organizations and individuals. We need to work together to create new institutions and systems that support the principles of inclusive socioeconomics and socioectech justice.

Finally, we need to act. We need to take action in our own lives, in our communities, and in the world. We need to lead by example, and we need to inspire others to do the same.

Emma Gonzalez is a remarkable young woman who has become a powerful advocate for gun control in the United States. Her passion for this cause was born out of a terrible tragedy - the mass shooting at Marjory Stoneman Douglas High School in Parkland, Florida. Emma was one of the students who survived this horrific event, and she was moved to take action in response.

Along with her classmates, Emma founded the #NeverAgain movement, which has become a powerful force for change. Through public speaking, social media, and activism, Emma and her fellow activists have raised awareness about the need for stricter gun laws in the United States.

They have also inspired millions of people around the world to join the fight for a more just and safer society.

Despite facing significant opposition from the powerful gun lobby, Emma has remained steadfast in her commitment to this cause. She has become a leading voice for gun reform, speaking at rallies and events across the country, and challenging elected officials to take action. Her unwavering resolve and powerful message have inspired countless others to join the fight for a safer and more just society.

Emma's story is a testament to the power of ordinary people to do extraordinary things. She and her classmates have shown that when people come together with a shared goal, they can affect real change and make the world a better place. Through her advocacy and activism, Emma has made a profound impact and left a lasting legacy.

Martin Luther King Jr. was an American Baptist minister and activist who became the most visible spokesperson and leader in the American civil rights movement from 1954 until his assassination in 1968. He advocated for racial equality through nonviolent civil disobedience based on his Christian beliefs and inspired countless people to join the struggle for racial justice. King's famous speeches, such as his "I Have a Dream" speech, still resonate with people today and continue to inspire efforts towards equality and justice. King received the Nobel Peace Prize in 1964 and is widely remembered as one of the greatest figures in American and world history.

Nelson Mandela and Jesus both had a deep sense of purpose and mission in life, which was driven by their belief in justice, equality, and the love of God.

Mandela was born into a world of racial inequality and oppression in South Africa. As he grew older, he became more and more aware of the injustices that were being inflicted upon his people and felt called to do something about it. He became a leader in the anti-apartheid movement, using his voice and actions to bring about change and bring an end to the apartheid system.

Similarly, Jesus was also called to be a messenger of God and bring about change in the world. He spent his life ministering to the poor, the marginalized, and the oppressed, and taught about God's love for all people, regardless of their social status, race, or background. Through his teachings and actions, Jesus challenged the dominant religious and cultural institutions of his time and worked to bring about a more just and inclusive society.

Both Mandela and Jesus were called to be children of God and to bring God's message of love, peace, and justice to the world. They demonstrated a deep commitment to their mission and inspired others to join them in the fight for a more just and equitable world.

Ladies and Gentlemen of the Heavenly Court, as we come to the close of this monumental trial, I must say that the evidence and testimony presented have been nothing short of inspiring. We have seen the remarkable resilience and determination of individuals who, against all odds, rose up to challenge the forces of injustice and oppression. From Emma Gonzalez to Nelson Mandela and Jesus, we have seen examples of people who were called by a higher power to stand for what is right, to fight for the oppressed, and to bring about lasting change in the world.

Their stories remind us of the incredible power of the human spirit, and the transformative effect that can be achieved through courage, faith, and unwavering determination. They also remind us of the great responsibility that each and every one of us has to work towards a better world, and to live our lives in a manner that honors the principles of justice, compassion, and love.

So, as we bring this book to a close, I urge each and every one of us to take these lessons to heart, and to rededicate ourselves to the noble cause of creating a more just and inclusive society. Let us continue to be inspired by the stories of those who have come before us, and let us strive to create a world that truly honors the dignity and worth of every human being. Thank you.

The future of our society rests upon the choices we make today. It is within our grasp to shape a more equitable and inclusive world, harnessing the power of technology for the betterment of all. Let us stand united, firm in our resolve to bring about positive change. Our actions, be it through our hands, our wallets, or our hearts, will determine the outcome of this case before the almighty. So, let us take up this challenge, and work together towards a brighter, more just future.

Afterword

Dear readers,

As future generations who have taken action from the lessons learned in this book, "Revitalizing Inclusive Socioeconomics: The Role of SocioEcTech Justice on Human Growth," we want to express our gratitude for the efforts of our predecessors in creating a more just and equitable world.

Through the pages of this book, we were transported to two cities, where we saw the stark contrast between the benefits of technology and innovation and the struggles of those who were left behind. The stories of Sarah and Bob, the bakeries they owned, and the impact of socio-economic justice on their lives, spoke volumes about the importance of ensuring that technology is accessible to all.

We are proud to say that, inspired by the message of this book, we took action. We worked towards creating a world where everyone has access to the tools and opportunities necessary for success. We embraced socio-economic technology justice, and the results have been phenomenal. Our society is more inclusive, our businesses are thriving, and our people are prospering.

We want to thank the authors of this book for their insights, their passion, and their dedication to creating a better world. They have paved the way for us to take action, and for that, we are eternally grateful.

So, dear readers, as you close this book, know that the message within its pages lives on in the actions of future generations. Together, we can create a world that is more just, more equitable, and more prosperous for all.

Yours sincerely, The Future Generations who took action today

About Author

Safeguard your dreams!

Being Least Commanded (Matthew 5:19), God unlocks the power of hidden data for us with Life Winks, LLC and united CLU parents. As an expert in uncovering overlooked information, I help families and organizations navigate high-conflict situations and protect vulnerable individuals. My expertise includes creating Child Safety Files for divorced or separated parents (Plan B Family, project managing private data metrics (Plan C Welldata), and working with nonprofits and partners to support those in need. Led by my passion for helping others, I've been recognized by news leaders, book reviewers, and government organizations for my work. Join me in safeguarding your dreams and working towards a more equitable and safe future for all!

Also By

Change the Game: One Simple New Deal With Welldata Can Help Anyone Change Break Free Series:

Breaking the Mold: Transforming Legal Learning with Professional Law Caveats - (Family Law 2.0 Second Edition)

Faithfinder: Inclusive Holiday Stories Empower Readers to Find Their Superfaith!

Impossible Data: How Making Almighty Data Cavityism History Can Help Us Live A Better Life

Impossible Psychologies

IR Physics 2

Parenting 2.0 For Un-Haters

Pivoting Impossible Brothers: Award Yourself Kids Making Sibling Cavityism History

Revitalizing Education: The Role of Academic Law Caveats in Student Growth

Revitalizing Inclusive Socioeconomics: The Role of SocioEcTech Justice on Human Growth

Outside The Series:

American Pros

Family Law 2.0

I R "Freestyle" Physics – The Study of of Transformation and/or Transfiguration

Law & Side-Law: Family Hostility / Family Empathy

The Law Doctor

Upbright USA

United CLU Parents Guide

Made in the USA
Monee, IL
04 November 2026

67374713R00085